MAKING AND UNMAKING OF THE SAN FRANCISCO BAY

MAKING AND UNMAKING OF THE SAN FRANCISCO BAY

Gary C. Howard and Matthew R. Kaser

CRC Press
Taylor & Francis Group
Boca Raton London New York

CRC Press is an imprint of the
Taylor & Francis Group, an **informa** business

First edition published 2021
by CRC Press
6000 Broken Sound Parkway NW, Suite 300, Boca Raton, FL 33487-2742

and by CRC Press
2 Park Square, Milton Park, Abingdon, Oxon, OX14 4RN

© 2021 Taylor & Francis Group, LLC

Library of Congress Cataloging-in-Publication Data

ISBN: 978-1-138-59672-6 (hbk)
ISBN: 978-0-367-74771-8 (pbk)
ISBN: 978-0-429-48749-1 (ebk)

Typeset in Times New Roman
by MPS Limited, Dehradun

Dedication

To Greg

Contents

Credits for Illustrations

Figure 1.1 **The Golden Gate.** The photograph is in the public domain. https://www.publicdomainpictures.net/en/view-image.php? image=212705&picture=golden-gate-bridge

Figure 1.2 **San Francisco Bay.** (*left*) Sketch by Gary Howard (GH). (*right*) Astronaut photograph ISS037-E-2604 was acquired on September 25, 2013, with a Nikon D3S digital camera using a 50-mm lens, and is provided by the ISS Crew Earth Observations experiment and Image Science & Analysis Laboratory, Johnson Space Center. The image was taken by the Expedition 37 crew. https://visibleearth.nasa.gov/ images/82198/san-francisco-region-at-night/82198f

Figure 2.1 **Farallon plate.** Sketch by GH.

Figure 2.2 **Subduction.** The illustration from USGS and in the public domain. https://www.usgs.gov/media/images/subduction-zone-graphic

Figure 2.3 **San Andreas Fault.** The illustration is by USGS and in the public domain. https://www.usgs.gov/media/images/san-andreas-fault-3

Figure 2.4 **Faults in the Bay Area.** Source: USGS at Earthquake outlook for the San Francisco Bay region 2014–2043 Fact Sheet 2016-3020

Figure 2.5 **Volcanic Activity in the East Bay.** Photograph by GH at Sibley Regional Park.

Figure 2.6 **Mount Diablo.** Photograph by GH from Sibley Regional Park.

Figure 2.7 **Track of the San Andreas Fault.** Photograph by Robert E. Wallace, USGS, and it in the public domain. https://pubs.usgs.gov/gip/ dynamic/San_Andreas.html

Figure 2.8 **Rocks of the Bay Area.** Photographs by GH at Sibley Regional Park, Don Edwards National Wildlife Refuge, and Lake Chabot Regional Park.

Figure 3.1 **Atmospheric Rivers.** The image is from the United States Naval Research Laboratory, Monterey, and is in the public domain. https://commons.wikimedia.org/wiki/File:Atmospheric_River_ GOES_WV_20101220.1200.goes11.vapor.x.pacus.x.jpg

Figure 3.2 **Lake Corcoran.** Sketch by GH.

Figure 3.3 **Carquinez Strait.** Photograph by GH.

1 California Now and Then

The story of the San Francisco Bay begins before there was a California or at least that part of the state that is west of the Sierra Nevada Mountains. It begins with the collision of two of the Earth's massive tectonic plates. The heavier oceanic Farallon plate was subducted by the lighter continental North American plate. As the Farallon plate dived under the other, considerable amounts of material from the floor of the Pacific Ocean were scrapped up and added to the growing coast of California. As the last portions of the Farallon plate were subducted, a few pieces broke off to open windows to part of the mantle that allowed very hot rock to reach the surface as small volcanoes in what is now the East and North Bay. Finally, with most of the Farallon plate out of the way, the Pacific plate collided with the North American plate to form the San Andreas fault.

The rocks that were scrapped up off the bottom of the ocean make up much of Bay Area and of Coastal California in general. As they were added to the North American plate, they broke off into distinct sections, called terranes. Terranes are masses of rock that were accreted from the Pacific Ocean floor to the growing coast of California. The terranes consist of heterogeneous rock. Their western boundary is the San Andreas fault. The most common rocks in this Franciscan Formation are a hard sandstone called graywacke, shales, and conglomerates that have all been partially metamorphosized. Others include chert, basalt, limestone, serpentinite, and blue schists. Outcrops of all of these can easily be seen in the Bay Area.

Part of the landmass that would eventually form the San Francisco peninsula began in Southern California. Part of the Southern California batholith on the western side of the San Andreas fault was detached from the mainland. In other words, it is part of the Pacific plate that is moving northward as it slips along against the North American plate. About 4.5 million years ago, that land mass had migrated northward to about the position of present-day Monterey, California. By about 700,000 years ago, the land mass had reached its current position.

The Bay was partially formed by a catastrophic event some 560,000 years ago. At that time, a large lake existed in the central valley of California in the valleys of the present-day Sacramento and San Joaquin Rivers. Its surface covered 12,000–19,000 square miles, and its southern end was originally open to the Pacific. Northward movement of the exposed parts of the Pacific plate along the San Andreas fault closed the opening to the sea and today extends into San Francisco. Further tectonic movements created a new opening to the sea through

1

FIGURE 1.1   The Golden Gate. The Gate is a narrow opening of less than a mile. The depth of the water there is about 350 feet, whereas the rest of the Bay is typically 10–20 feet.

the Carquinez Straight. The fast-moving water carved out canyons to form the northern part of San Francisco Bay and the Golden Gate. For much of its existence, the Bay was a large valley with rivers that flowed out through the Golden Gate to the Pacific Ocean (Fig. 1.1).

In fact, the Bay had alternated between wet and dry over the last 2 million years as Ice Ages came and went and sea levels rose and fell. Its evolution is a fascinating story. It involves the movements of tectonic plates, the rise and fall of oceans, the coming and going of water, and the actions of humans. It was not always here, and it will not be here it the future. Changes in the global climate resulting from the last ice age yielded lower and then higher ocean levels.

Approximately 10,000 years ago, Native Americans living in the San Francisco Bay Area would have noticed changes to their environment. Year after year, the tides were higher than those before. The ocean was creeping further and further onto the land. Eventually, the water poured between the two great hills that formed the Golden Gate and began to fill the valley floor. The Bay was beginning (Fig. 1.2).

The San Francisco Bay encompasses about 1600 square miles and drains about 75,000 square miles or nearly a quarter of California. It connects to the salt water of the Pacific Ocean through a narrow opening called the Golden Gate. Several rivers empty fresh water into the Bay, including the Sacramento and San Joaquin, Napa, Petaluma, and Guadalupe Rivers and a number of small streams

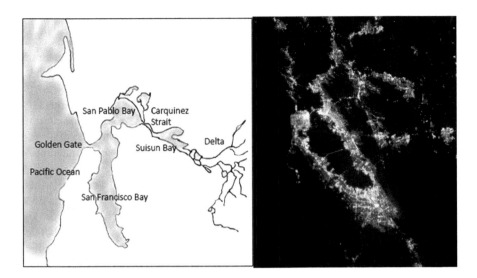

FIGURE 1.2   San Francisco Bay. (Left) The Bay is a large estuary and includes the San Francisco Bay, San Pablo Bay, and Suisun Bay. Fresh water enters mainly through the Sacramento River Delta and mixes with salty seawater entering though the Golden Gate forming gradients of salinity that change with the seasons and the tides. The Carquinez Strait is also a narrow passage that was carved by a massive flood in about 640,000–700,000 years ago. (Right) The Bay Area is one of the largest metropolitan areas in the United States and includes the cities of San Francisco, Oakland, San Jose, and many more.

also flow into the Bay at least intermittently. Other watersheds, marshes, and creeks also feed the Bay. The Bay is the largest estuary on the west coast of North and South America. As an estuary, the Bay is influenced both by the upwellings and movements of the Pacific Ocean at its mouth and the addition of fresh water from the rivers and streams that flow into it. Winters in Northern California are wet and summers are dry, and the amount of runoff water varies greatly from a low of $7.6\,\text{km}^3$ in 1977 to a high of $65\,\text{km}^3$ in 1983 (Cloern and Jassby, 2012). The average annual runoff is somewhere between 20 and $35\,\text{km}^3$. Much of the water comes in the form of snow in the Sierra Nevada Mountains to the east. In the summer, the snow melts and flows into the rivers and streams of the watershed. The watershed provides drinking water for 25 million Californians and irrigation water for 7000 square miles of agricultural land.

The Bay is nearly surrounded by land. Only the narrow Golden Gate (about 1 mile) is open to the Pacific Ocean. San Francisco sits at the northern tip of a long peninsula at the southern end of Golden Gate. At the northern end is Marin County. Oakland and Berkeley are on the eastern side of the Bay, and San Jose is at the southern end. The Bay itself consists of three large bays. The main bay is San Francisco Bay, which is surrounded by San Francisco, Oakland, and San Jose. San Pablo Bay is to the north and beyond it is Suisun Bay. The Sacramento-San Joachin River Delta empties into Suisun Bay. It is the only inland delta in the world. The Delta began as a plain covering about 1000 square miles. Periodic flooding and tidal movements produced a brackish marsh and a maze of branching channels. Other smaller bays and wetlands surround much of the Bay. The wetlands have been greatly reduced since the arrival of Europeans to the Bay Area, but efforts are being made to replace them now.

While water is not quite so powerful a force as the tectonic plate movement, it has nevertheless had a significant role in forming the Bay Area. The rupture of Lake Corcoran released a torrent of water that carved the Carquinez Strait between San Pablo and Suisun Bays. Rising and lowering sea levels have also influenced the size of the Bay itself.

The Bay has a diverse ecology and has been home to a wide variety of plants and animals since its formation. Over the millions of years, those plants and animals have evolved. Some have become extinct, and others have developed to take their place. Early on the animals included many of the large mammals that characterize western North American. The Bay Area is also part of the Pacific flyway for migratory birds, and in 2012, it was recognized by the Ramsar Convention on Wetlands of International Importance for waterfowl habitat, an international treaty for conserving wetlands.

Human activity influenced the Bay. Hydraulic gold mining in during the California gold rush sent masses of slit into the Bay. Much of the silt from 19th-century hydraulic gold mining has been washed out, but it still affects Bay plants and animals today. Humans have also built several major cities and filled significant parts of the Bay. The Bay is quite shallow over most of its area. In the 19th century, settlers began building levees to protect their land, and now there are more than 1100 miles of levees protecting farms and residential areas. Those levees have also greatly affected the movement of water in the area and thus the plants and animals there. The average depth is 12–15 feet. For many years, the Army Corps of Engineers has overseen dredging of the shipping channels that allows ships to access the ports of Oakland and Sacramento. Vessels with a 50-foot draft can be accommodated at Oakland.

Today, the San Francisco Bay Area comprises about 7000 square miles of hills, valleys, and an enormous system of interconnected bays. The Bay itself is a shallow estuary that is fed by the Sacramento, San Joaquin, and Napa Rivers. It is surrounded by a large population center. More than 7 million people live in the Bay Area, and it is one of the largest metropolitan areas in the United States.

The San Francisco Bay and the area around it is beautiful and dynamic place. It is one of the most populous areas in the United States with three major cities, a huge natural harbor for shipping, and a destination spot for tourists from around the world. The Bay was not always here, and the geological forces that created it are still active and will one day destroy the Bay. Here we will attempt to describe the Bay in the past, present, and future, and to explain those forces.

The Bay Area is home to about 7.75 million people. It includes nine counties (i.e., Alameda, Contra Costa, Marin, Napa, San Mateo, Santa Clara, Solano, Sonoma, and San Francisco) and the cities of San Francisco, Oakland, San Jose, and numerous smaller ones. The Bay Area continues to increase in population, and the additional people put more stress on the housing, transportation, and other systems in the Area. Eight major bridges cross the bays and rivers to connect the cities.

Six major faults cross the Bay Area, and thus, it is and has been subjected to considerable tectonic activity over the years. Earthquakes and ground movement are the major forces that built the Bay, and periodic earthquakes always will rock the Bay Area. The largest most recent earthquakes were the Great San Francisco earthquake and resulting fire in 1906 that destroyed most of the city and the Loma Prieta earthquake of 1989. Both of those were on the region's largest fault, the San Andreas fault. The Hayward fault in the East Bay is nearly as dangerous and also overdue for a large earthquake.

Movement on those faults is what built the Bay and what will eventually destroy it. Other forces include the rise and fall of the ocean with changes in climate, atmospheric rivers, and finally humans. In the crush of our daily activities, it's easy to forget that these forces are not influenced by our meetings, jobs, smartphones, or desires. They move at their own speed and come in their own time. We are now feeling the effects of the climate crisis with much hotter temperatures and seemingly endless fire seasons. The sea level is beginning to rise as the Earth's large icecaps and glaciers melt. That will change the coast of the ocean and Bay over the next years. Earthquakes periodically rumble through the Bay Area to remind us that we are still part of the natural world.

The forces that built it began with plate tectonics and the movement of the plates that formed the western part of modern California. These involved the collision of the Pacific and North American plates and the subduction of the Juan de Fuca plate. Thus, the Bay Area has marine but not terrestrial dinosaur fossils.

This book will describe the natural history and evolution of the San Francisco Bay Area (e.g., its geology, plants and animals, people and their activities, and the connections among these) over the last 50 million years through the present and into the future. We will explain how those forces that built the Bay are still active. We will review the predictions that scientists have made about the future of the Bay Area and how those forces will eventually change the Bay beyond recognition. Humans witnessed the end of the Ice Age, the rise of sea levels, and the flooding of the Bay. We do not know if humans will see the end of the Bay, but the end will come.

REFERENCES

Cloern J.E., Jassby A.D. (2012) Drivers of change in estuarine-coastal ecosystems: discoveries from four decades of study in San Francisco Bay. *Reviews of Geophysics* 4: 395–408.

2 Geological Forces that Built the Bay

The story of the San Francisco Bay Area begins deep in the Pacific Ocean about 165 million years ago, roughly in the early Late Jurassic. Pangaea had just broken into two supercontinents, and the Atlantic Ocean was forming in the gap between them. The continents were ruled by dinosaurs, such as the *Brontosaurus* and *Stegosaurus*, and the largest predator of all at the time, *Allosaurus*. But there are no terrestrial dinosaur fossils to find in the Bay Area, which comprises igneous, marine and coastal sedimentary, and metamorphic rocks from this time. None of these giant creatures roamed the Bay Area. They became extinct well before the West Coast of California formed.

SUBDUCTION

The land mass that would later become the Bay Area resulted from titanic forces that have and continue to reshape the continents. Those forces have dramatically changed the Earth's surface. The Earth has three layers, somewhat like an egg. The yolk represents the extremely hot semi-molten core. The white is similar to the mantle. And the egg shell is similar to the crust. The crust forms only the top approximately 18 miles on the continents and about 3 miles under the oceans. It is divided into massive plates that "float" on the more dense material of the mantle far below the surface. The mantel and core are still extremely hot, and as regions of hot mantle rise to the surface, they pour out material that forms new crust. The new material pushes aside the older material and causes the plates to move about over the Earth's surface. During those movements, the plates often collide with one another in different ways. They might run directly into each other to produce a crumpled boundary that forms large mountain ranges, such as the Himalayas. In other cases, such as the San Andreas fault, the plates slip along each other. Finally, one plate might ride up over another. Oceanic plates tend to be more dense than continental plates, and so, they tend to push their way underneath the continental plate (Fig. 2.1).

That last case is just what happened to form California. Much of the material that would eventually form that California coast was part of the Farallon plate, which formed the floor of part of the Pacific Ocean. As the Farallon plate migrated eastward, it bumped into the North American plate and, being more dense, subducted under the North American plate. The friction of the two plates moving against each other caused the subducted material to melt. The resulting plumes of melted rock caused volcanic activity that formed the volcanoes in present-day Eastern California. Today, the Farallon plate has been entirely subducted

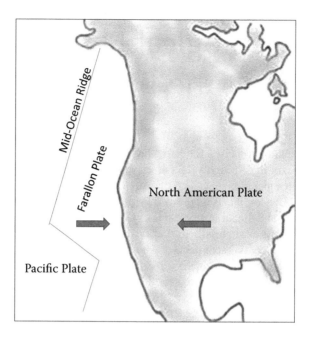

FIGURE 2.1 Farallon Plate. The making of the Bay Area involved the enormous forces that move the Earth's tectonic plates around the surface. Spreading of the mid-oceanic ridge pushed the Farallon plate toward the east so that it collided with the North American plate, which was moving west.

underneath the North American plate. However, its influence has been felt several times over the ages. For example, its movement and the resulting heat were responsible for the massive earthquakes on the New Madrid fault in the Mississippi River Valley in the early 19th century (Forte et al., 2007). In addition, subduction of a fragment of the Farallon plate, the Juan de Fuca plate, is providing the energy for the Cascadian volcanoes.

Even 65 million years ago, the west coast of California was in the Sierra foothills. The rest of California was still under the Pacific Ocean. As the subduction continued, more material was scraped up onto the edge of the continent. About 25 million years ago, the subduction changed to a strike-slip boundary, and marine sedimentary material settled into the trench left by the earlier subduction. These are represented in the Monterey Formation on the current Monterey Peninsula and Claremont Formation in the East Bay (Fig. 2.2).

The change in the nature of the boundary between the plates also resulted in a series of small volcanoes in the Bay Area. The energy that powered these volcanoes did not come about from the subduction of a plate, such as those in present day eastern California. The movement and opening of faults in the area allowed magma to move upwards and flow out to yield many of the formations from the South Bay through the East Bay. Those same movements along the

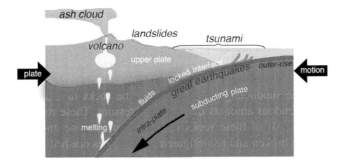

FIGURE 2.2 Subduction. Over many years, the heavier oceanic Farallon plate was subducted under the lighter continental North American plate. Material from the ocean floor was scrapped up onto the North American plate to form most of what is now the coastal pieces of California. Most of the Farallon plate is now underneath the North American plate. All that remains are a few small pieces, such as the Juan de Fuca plate off the Oregon and Washington coasts (Illustration from USGS).

faults also moved blocks further north. For example, the dead volcanoes in the Berkeley Hills were active about 20–30 million years ago when they were closer to San Jose.

The Franciscan complex in the Bay Area can be subdivided into nine terranes (Elder, 2013). The terranes are large pieces of material from the seafloor of the Pacific Ocean that were scrapped up as the Pacific plate was subducted by the North American plate over 100 million years. Each of the terranes has a different composition of rocks. Those collections of material formed the current California West Coast. The somewhat random material is sometimes called a mélange. The subduction causes enormous friction between the moving plates, and that heat causes volcanos to form inland. The rocks of the mélange form in from different processes. In the deep ocean, dead planktic material and clay particles form a deposit on the crust. Overtime, the carbonate components dissolve, leaving mostly silica, and that silica and mud develop into the ribbon chert that is common in the Bay Area. In more shallow areas, the water is warmer, and organisms thrive. Their deposits tend to be rich in carbonates and form limestone. Sandstone and shale are eroded from the continental plate onto the accumulating ocean floor material. The uplifted slabs form an accretionary prism in which the older rocks are on the eastern side. Volcanic action from the subduction provides additional types of metamorphic and igneous rock to the mélange.

From about 165 to 40 million years ago, this process was actively building the current coast of California, and the rocks that form the Bay Area are now known as the Franciscan complex. As more and more material collected, the California coast continued to move westward. Much of that rock originated as pillow basalt that erupted thousands of miles out in the Pacific Ocean from 100 to 200 million years ago. That material entered the "conveyor belt" that moved it from its origin to the coast over millions of years. The complex also includes chert that formed from the countless numbers of microorganisms called radiolarians. Once they die, their silica shells fall to the ocean floor and harden into chert. Some of the deposits in the Bay Area are 250 feet thick. Other components of the Franciscan complex include sandstone and shale. These sedimentary rocks come from the North American plate and were included in the mix when sections of continental crust fell onto the subducting plate. Some of the rocks in the complex were subjected to tremendous amounts of heat and pressure. These rocks changed or metamorphosized. All of these rocks formed layers on one another, and then those blocks were broken and reconfigured as the plates crashed into each other.

SAN ANDREAS FAULT

Then, about 30 million years ago, things changed as the Farallon plate was nearly completely subducted under the North American plate. That left the Pacific plate in contact with the North American plate, and the boundary between the two plates formed the San Andreas fault. The fault was found by Andrew Lawson (University of California, Berkeley) in 1895. The northern segment runs from Hollister through the Santa Cruz Mountains and up the peninsula to San Francisco. It forms the boundary between the Pacific plate and the North American plate. The Pacific plate is moving roughly northwest, and the North American plate is moving westward. The motion along the fault is technically called a right-lateral strike-slip fault. The fault formed about 30 million years ago. It can easily be seen from the air. For example, the fault lies in the bottom of the string of lakes, Crystal Springs Lakes, on the San Francisco Peninsula and can be tracked further south by just looking at the landscape (Fig. 2.3).

The San Andreas fault is actually part of a system of faults throughout the Bay Area. Running along the coast are the San Andreas and San Gregorio faults. In the East Bay, a series of faults include the Hayward, Calaveras, Rogers Creek, and several others. All of them join just south of San Jose. Earthquakes occur periodically on all of the faults as the stress from the movement of the plates builds up. After it reaches some threshold, the energy is released as the plates slip past each other to a new position. Most earthquakes are very small, but some quite large earthquakes have taken place in the last 100 or so years. In the San Francisco earthquake of 1906, some parts of the Bay Area moved 21 feet. The more recent Loma Prieta earthquake of 1989 resulted in about 6 feet of movement. Those were both on the San Andreas fault. The Hayward fault tends to creep along at about ¼ inch per year. Nevertheless, large earthquakes can still occur on the East Bay faults. Why some faults rupture and others creep is not understood. Some have suggested that the common mineral serpentine might be involved. It is a soft and slippery material, but no serpentine has been found on the Hayward Fault (Fig. 2.4).

As the San Andreas fault formed, it cut across the subducting Farallon plate and formed an opening for magma to flow upward (McCrory et al., 2009). The result of this "slab window" was three volcanic events in the Bay Area. About 11–8.5 million years ago, the first event covered the area from Hollister through

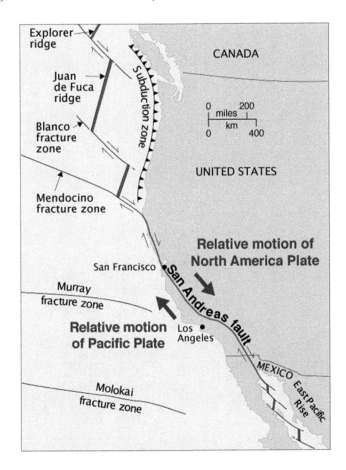

FIGURE 2.3 San Andreas Fault. The San Andreas fault marks the current line of collision between the North American and Pacific Ocean plates along the California coast. The North American plate is moving southeast, and the Pacific plate is moving northwest. The plates are slipping along past one another, but occasionally, the plates become stuck at a point. Stress builds up and is released as an earthquake. Over millions of years, the portion of California on the Pacific plate has moved northward and is continuing that movement even today (Illustration is by USGS).

the East Bay Hills, part of Marin County, and on up to Petaluma and Santa Rosa. The second event occurred 8–2.5 million years ago and left deposits in Sonoma County. The third took place from 2.5 million to 10,000 years ago and is still somewhat active in the Clear Lake and Geysers region. Mount Konocti is a prominent feature of this volcanic activity. It first erupted about 350,000 years ago, and its last eruption was about 11,000 years ago. The Geysers with their famous hot springs are heated by the remnants of these volcanoes. Evidence from recent earthquakes indicates that this area is still active (Hayes et al., 2006).

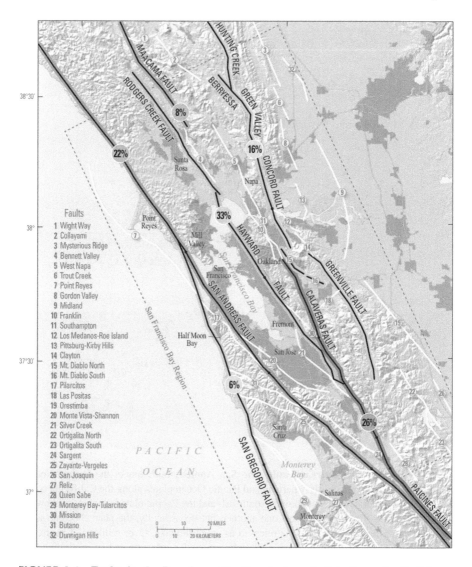

FIGURE 2.4   Faults in the Bay Area. The San Andreas fault is the major fault in California. However, in the Bay Area, it is actually a system of faults that include the Hayward, Calavaras, San Grigorio, Concord-Green Valley, Greenville, Rodgers Creek, and several more minor faults. Map courtesy of USGS.

There are other examples of movement on the San Andreas fault. One of the most impressive is the Pinnacles, which are located today just east of Soledad. The Neenach volcano erupted 23 million years ago. At that time, it was near the city of Lancaster nearly 200 miles south of its current position. The volcano lies on the fault, and as the Pacific plate has moved northward, half of the volcano moved with it, and the other half remained in place. The columns that form the Pinnacles represent the solidified magma of the volcano after the softer rock that surrounded the columns has been eroded away. In addition, the main trace of the San Andreas fault has move 4 miles to the east.

About 10 million years ago, the Moraga Volcanics, now in the hills above Hayward and Oakland in the East Bay but then southeast of San Jose, spewed out large amounts of lava (Case, 1968) from one large and three smaller volcanoes. The lava flowed out across an alluvial plain that had once been covered by water. Since then, the large volcano exploded, and its remnants are under the site of the present-day Lawrence Hall of Science in Berkeley. The movements of the Hayward fault have caused another, called Round Top, to fall over on its side and moved all of the volcanic rock formation northward to their present site.

The Coast Ranges in the North Bay are a geological puzzle of Mesozoic and early Tertiary rocks that resulted from the collision of two continental plates (Wagner et al., 2011). These have since been overlain with sedimentary and volcanic rocks. Interpreting them is challenging. The volcanic activity was produced by a slab window that opened up to allow material from the mantle to migrate upward. These occurred throughout much of the East Bay northward into the area of the Sonoma and Clear Lake Volcanics and southward to the Tolay and Berkeley Hills Volcanics and much further south to the Quien Sabe Volcanics near south of modern-day San Jose. However, the volcanic activity was much greater in the more northern areas.

Over the last 12 million years, the San Andreas fault has moved 175 km (Langenheim et al., 2010). That movement produced the Bay of today, and the San Andreas and the other faults of its system contributed to varying degrees. Transpression caused the land to rise to form hills, and transtension caused the land to droop to form the valleys and basins. The Napa and Sonoma Valleys are well known for their grape and wine production. Those basins were formed by a combination of the strike-slip movement on the faults, folding, and the actions of volcanoes. The Santa Rosa area formed under transtension that later turned to transpression. The multiple forces that have acted on these areas are not unusual. The activities have been ongoing for millions of years. In fact, they will continue to modify the land structures with movement laterally and vertically indefinitely into the future.

Although the geology of the Bay Area is complicated, it can be looked at as a collection of blocks that are each surrounded by faults (D'Alessio et al., 2005). The blocks have moved together as a unit for millions of years, and they have been moving generally toward the northwest. The smaller blocks of the Bay Area can be assembled into three large blocks. The Salinian block is west of the San Andreas fault and is attached to the Pacific plate. The region between the San Andreas and Hayward faults is the San Francisco block, and the area east of the Hayward fault is the East Bay block. The latter two blocks are on the North American plate and also generally moving northwest, but at different rates.

The fault system in the East Bay is complicated. Two of the main faults are the Hayward and the Rodgers Creek faults. For some time, it was not clear if those two were, in fact, a single fault. The region between them lies at the bottom of the San Pablo Bay. That question was solved by a recent study involving subsurface imaging (Watt et al., 2016) that definitively showed that the two are connected. While that finding has interest for geologists, it has practical implications for the people who live and work along that fault system. The larger fault means that the size of an earthquake can be much greater.

Some sections of the fault system in the Bay Area continue to creep as they have for some time. They also tend to have repeating small earthquakes, and those earthquakes can be used to characterize the faults. Those creeping faults are less likely to build up large amounts of energy and release it suddenly when the fault ruptures. A study by Shakibay Senobari and Funning (2019) found that three faults (Maacama, Rodgers Creek, and Bartlett Springs) are creeping and, thus, less likely to have a serious earthquake.

About 3.5 million years ago, the nature of the interaction of the Pacific and North American plates changed. They began to partially oppose one another rather than simply slide along each other. This combination of strike-slip and compression is called transpression (Dewey et al., 1998), and that compression squeezed the land mass of the Bay Area, causing some of the hills to rise up. Most of the hills in the Bay Area, including Mount Diablo, the highest peak in the Area, are a result of this compression. For the last couple million years, the intersection of the two plates returned to a strike-slip mode, and the uplift of the land due to transpression ceased.

Other forces have continued to cause some areas of land to rise up. The faults are not straight lines. They have bends and curves. Movement at those points causes the land surface to be pushed up or to be depressed, depending on the direction of the bend. Loma Prieta, Mount Umunhum, and Mission Peak resulted from this type of uplift.

VISIBLE REMINDERS OF THE FORCES THAT BUILT THE BAY

It has taken over 100 million years for the geologic forces to make the San Francisco Bay Area. The Bay Area was formed by the subduction of the heavy Farallon oceanic plate underneath the lighter North American continental plate. The remnants of the Farallon plate broke off to allow windows of hot magma to flow upwards and add to the sedimentary rocks from the two plates. Eventually, the entire Farallon plate was subsumed beneath the North American plate, and that allowed the Pacific plate to collide with the North American plate at a strike-slip boundary that became the San Andreas fault system. Although the geologic forces that made the Bay Area are still just as active, they move at a rate that is far longer than the ability of human observation to perceive them (with the exception of the occasional earthquake). Thus, as far as we humans know, the Bay has always been here and will always be here.

Still, evidence of the geological past can easily be seen in the Bay Area today (Fig. 2.5).

FIGURE 2.5   Volcanic Activity in the East Bay. Areas of volcanic activity are seen throughout the East and North Bay. This photograph shows a sequence of volcanic layers. Beginning at the left is dark gray basaltic lava. Basaltic tuffs or ash landed on the still hot lava and changed colors to orange and pink. The whole complex has been tilted up to nearly vertical. In addition, over the last millions of years, this whole area has moved northward.

Moraga Volcanics

The East Bay hills feature a group of extinct volcanoes called the Moraga Formation or Volcanics. The largest volcano is essentially gone now. Remnants of smaller volcanos are found south of Berkeley. One called Round Top is in Sibley Volcanic Regional Preserve. Further folding of the East Bay tilted Round Top onto its side, and quarry operations exposed the volcano's structure. The volcanos erupted about 10 million years ago, allowing lava to spread over the sedimentary rocks that had been laid down by erosion of material in the Coastal Range to the west. The burned sedimentary rock can easily be seen close to the basalt flows near Round Top.

The energy for these small volcanoes was provided by slab windows that opened up as the Farallon plate was being subducted under the North American plate. As the slabs broke off from the Farallon plate, they allowed hot material from the mantle to well up so that magma rose to the surface (Fig. 2.6).

FIGURE 2.6   Mount Diablo (3849 feet), the Highest Mountain in the Bay Area, Looks Somewhat like a Volcano, it is Not One. It resulted from geologic compression and uplift and continues to grow at a rate of 3–5 mm per year. The mountain looks like two, but actually, it has two peaks. The top includes volcanic and sedimentary rocks from the Farallon plate 90–190 million years ago. Like much of the rock in the Bay Area, the higher parts of Mount Diablo consist of rock that was scrapped up from the floor of the Pacific Ocean as the Farallon plate was subducted under the North American plate some 90–190 million years ago. Those and other volcanic and sedimentary deposits from the Franciscan complex and Mount Diablo Ophiloite make up the rest of the mountain. Over many millions of years, the area has been changed by compression, folding, buckling, and erosion.

MOUNT DIABLO

At 3849 feet, the highest peak in the Bay Area is Mount Diablo. It resulted from compression of the boundary between the Pacific and North American plates. The mountain was pushed up by a thrust fault on the southwest flank and con- tinues to be pushed up at a rate of 3–5 mm per year. The mountain includes rocks of many different types. Near the summit are rocks that were under the Pacific Ocean during the Jurassic and Cretaceous eras. Near the top are equally old various volcanic and sedimentary deposits. These rocks were from the sea floor above the Farallon plate and were scraped up onto the North American plate as the Farallon plate was being subducted under the North American plate. To the west of the mountain, the rocks are those of the Franciscan complex. To the east, there are sedimentary rocks (e.g., sandstone, mudstone, and limestone) that form the Great Valley Sequence. The mountain has been subjected to the same forces as the rest of the Coastal Range, including the actions of the San Andreas fault system, compression, folding, buckling, and erosion over the last several million years. Schemmann et al. (2007) provide an excellent description of the geologic actions on Mount Diablo.

CARQUINEZ STRAIT

The Carquinez Strait connects Suisun Bay and the Delta with San Pablo Bay. Long ago, much of the Central Valley was covered by a large prehistoric lake. Uplift of the California Coast Ranges blocked the run-off of water from the Sierra Nevada. About 560,000 years ago, the massive amount of water in the lake was released, and the rushing water carved a new path to the Pacific Ocean through the Carquinez Strait. The Strait is 1.0–1.5 km wide and about 14 km long. It features strong ebb tides and seasonal floods from the Sacramento and San Joaquin Rivers. Woodrow et al. (2017) examined core samples taken from the Strait. The base of the north wall features inclined strata with couplets of fine/medium grained sand alternated with mud. These may have resulted from accretion onto the wall. The top of the wall contains horizontal strata of couplets of silty clay and fine sand. A small number of cross strata were also seen. Folded and broken strata appear to be the result of soft-sediment deformation. Some fossils were found in the cores. These include bivalve shells, some trace fossils, and some plant material.

SACRAMENTO DELTA

The Delta is a product of the end of the last Ice Age. About 10,000 years ago, rising sea levels forced water back up through the San Francisco Bay and slowed down the flow of water from the San Joaquin and Sacramento Rivers. The narrow Carquinez Strait and tidal actions caused sediments to collect in the rivers. Early on, the Delta was a constantly changing collection of shifting channels, sand dunes, alluvial fans, and floodplains that were influenced by the rising sea levels. About 8000 years ago, the sea level rise slowed, and plants began to grow on the islands that changed character to peat and tule that further slowed the water flow. Now the Delta is characterized by shallow channels and sloughs surrounding those tule islands. The Delta islands continue to change. Sometimes that change can occur suddenly. For example, earthquakes and floods can radically change the lay of the islands (Mount and Twiss, 2005).

SONOMA VOLCANICS

These volcanic formations in Napa and Sonoma counties include most of the mountains in the two counties and are the source for the rich soils in the major wine-growing regions of California. They include the Quien Sabe Volcanics, the volcanoes in the East Bay, those in Sonoma County, and all the way up to the Clear Lake Volcanics (Sweetkind et al., 2011). They are several thousand feet thick. They were formed by eruptions from several vents more than 1 million years ago. Langenheim et al. (2006) used gravity measurements to determine the distribution and types of rocks in this area. The rocks are younger towards the northwest, and they have been disturbed by the various faults.

GEOTHERMAL SPRINGS

Until about 10,000 years ago, volcanos were active in the Bay Area. Those volcanos in the Bay Area are all extinct and quiet now. But in one area north of San Francisco, evidence of the previous volcanic activity can still be actively seen. The geothermal springs in Napa and Lake counties continue to regularly spray hot water and steam into the air, and fumaroles release hydrogen sulfide, which smells like rotten eggs. For example, the Geysers in the Mayaamas Mountains near Healdsburg and Geyserville are remnants of the eruptions many years ago. As recently as 2 million years ago, magma intrusions moved nearer to the surface (Erkan et al., 2005). In fact, models of the area suggest that magma intrusions must have occurred much more recently to maintain the high temperatures underground. At 3 km underground, the rock temperatures are still at 250°C–300°C. The energy comes from a very shallow magma chamber that is 5–10 miles below ground. Ground water seeps down through the rock through tiny fractures in the rock. As it gets closer to the rocks near the superheated magma, the water gets hotter and hotter until it changes to steam and forces its way back to the surface as a geyser.

The geothermal system at the Geysers results from an essentially closed system. The Mercuryville and Collayomi faults bound the system, and much of the rock in the faults is serpentinite, which is a soft and pliable rock that acts to seal the system. The rock in the system is a very hard form of metamorphosed sandstone called graywacke. However, it is riddled with microfractures that allow the cool ground water to seep downwards and then to race to the surface at 180°F.

The warm mineral-rich waters have been used for healing and relaxation, beginning with the Native Americans hundreds of years ago. That use continues today, but the area is also used to produce electricity. Water, steam, and hydrogen sulfide are not the only materials that move through the fractures in the rock, but gold, silver, and cinnabar (an ore of mercury) also collect in the fractures (Fig. 2.7).

TRACE OF SAN ANDREAS FAULT

The trace of the fault is easily seen in Tomales Bay. The Bay is a long narrow inlet, approximately 1 mile wide and 15 miles long. It can also be seen in the San Andreas Reservoir and along the ground further south.

COMPLEX GEOLOGY OF THE EAST BAY

The geology of the Bay Area is complex. It was formed from a combination of material scraped up from the bottom of the Pacific Ocean as the Farallon plate

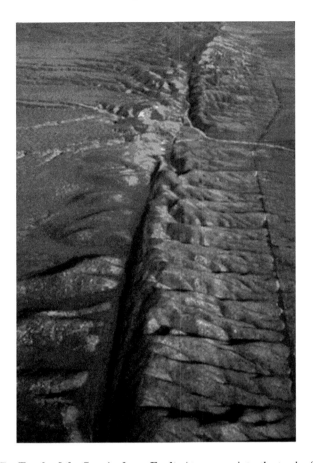

FIGURE 2.7   Track of the San Andreas Fault. At some points, the track of the fault is quite obvious. This section is just south of San Jose. The fault also runs through the bottom of the Crystal Springs Reservoir in the Santa Cruz Mountains just south of San Francisco and also through the Tomales Bay north of San Francisco. In this image, the fault can easily be seen on the Carrizo Plain just east of San Luis Obispo (Photograph by Robert E. Wallace, USGS).

was subducted under the North American plate, sedimentary material from the continental plate, and finally volcanic rock produced by the heat of the collision of the two plates. That collision further bent and fractured the different rocks. Various aspect of the material can be seen throughout the Bay Area. One example is at the east side of the Caldicott Tunnel in Oakland. From west to east, the different zones include Leona rhyolite, sandstone and shale, chert, conglomerate and sand, volcanic material, shale and limestone, and more rhyolite and sediments (Fig. 2.8).

FIGURE 2.8   Rocks of the Bay Area. In many places, rock outcrops tell an interesting story about the Bay Area. Much of the rocky material of the Bay Area was scrapped up off of the bottom of the Pacific Ocean as the Farallon plate was subducted under the North American plate. Other sedimentary rock was formed on the North American plate and added to the mix. (A) Chert, (B) ribbon chert, (C) green serpentine a semi-metamorphosized rock that was squeezed up like toothpaste by the earth movements, (D) volcanic activity often resulted in basalt intrusions, (E) mudstone, and (F) layers of sandstone.

SAND DUNES IN SAN FRANCISCO

Much of current San Francisco was covered with sand dunes. This include most of what is now Golden Gate Park. Interestingly, the source of the sand was the Sierra Nevada Mountains. During the last Ice Ages about 20,000 years ago, ice covered much of the Sierra, and the ice ground down the mountain rocks. The smallest particles were then carried down the rivers and through the Carquinez Strait to the Pacific Ocean where they were deposited. Over time, the sand was then blown onto the land by the westerly winds and eventually covered the San Francisco peninsula and even much of Oakland. The sand can still be found throughout San Francisco, especially in the Park.

ISLANDS IN THE BAY

The islands in the Bay are all the tops of hills that existed on the plain before the water returned to the Bay. Only Treasure Island is an artificial island. The largest island is Angel Island. At one time, it was attached to the Marin headlands, but water erosion eventually cut through the land connection to form Raccoon Strait and separate it as an island. Alcatraz is entirely made of the very hard sandstone called graywacke.

REFERENCES

Case J.E. (1968) *Upper Cretaceous and Lower Tertiary Rocks*, Berkeley and San Leandro Hills, California. US Geological Survey Bulletin 1251-J.

D'Alessio M.A., Johanson I.A., Burgmann R., Schmidt D., Murray M.H. (2005) Slicing up the San Francisco Bay Area: Block kinematics and fault slip rates from GPS-derived surface velocities, *Journal of Geophysics and Research* 110: B06403.

Dewey J.E., JR, Holdsworth R.E., Strachan R.A. (1998) Transpression and transtension zones. In: *Continental Transpression and Transtensional Tectonics*, edited by Dewey J.E., JR, Holdsworth R.E., Strachan R.A. (eds), Geological Society, London, Special Publications, 135: 1–14.

Elder W.P. (2013) Bedrock geology of the San Francisco Bay Area: A local sediment source for bay and coastal systems. *Marine Geology* 345: 18–30.

Erkan K., Blackwell D.D., Leidig M. (2005) Crustal Thermal Regime at The Geysers/ Clear Lake Area, California. *Proceedings World Geothermal Congress 2005*, Antalya, Turkey.

Forte A.M., Mitrovica J.X., Moucha R., Simmons N.A., Grand S.P. (2007) Descent of the ancient Farallon slab drives localized mantle flow below the New Madrid seismic zone. *Geophysical Research Letters* 34: L04308.

Graymer R.W. (2018) Overview of the geology of the San Francisco Bay Region. In: *Geology of San Francisco, California*, edited by K.A. Johnson, G.W. Bartow. Brunswick, Ohio, Association of Engineering Geologists. https://www. aegweb.org/assets/docs/updated_final_geology_of_san.pdf.

Hawley W.B., Allen R.M. (2019) The fragmented death of the Farallon plate. *Geophysical Research Letters* 46: 7386–7394.

Hayes G.P., Johnson C.B., Furlong K.P. (2006) Evidence for melt injection in the crust of northern California? *Earth and Planetary Science Letters* 248: 638–649.

Langenheim V.E., Graymer R.W., Jachens R.C., McLauglin R.J., Wagner D.L., Sweetkind D.S. (2010) Geophysical framework of the northern San Francisco Bay region, California. *Geosphere* 6: 594–620.

Langenheim V.E., Roberts C.S., McCabe C.A., McPhee D.K., Tilden J.E., Jachens R.C. (2006) Preliminary isostatic gravity map of Sonoma volcanic field and vicinity, Sonoma and Napa Counties, California. USGS Open-File Report 2006–1056.

McCrory P.A., Wilson D.S., Stanley R.G. (2009) Continuing evolution of the Pacific–Juan de Fuca–North America slab window system—A trench–ridge–transform example from the Pacific Rim. *Tectonophysics* 464: 30–42.

Mount J., Twiss R. (2005) Subsidence, sea level rise, seismicity in the Sacramento-San Joaquin Delta. *San Francisco Estuary and Watershed Science* 3: Issue 1, Article 5. https://escholarship.org/content/qt4k44725p/qt4k44725p.pdf.

Schemmann K., Unruh J.R., Moore E.M. (2007) Kinematics of Franciscan Complex exhumation: New insights from the geology of Mount Diablo, California. *GSA Bulletin* 120: 543–555.

Shakibay Senobari N., Funning G.J. (2019) Widespread fault creep in the northern San Francisco Bay Area revealed by multi-station cluster detection of repeating earthquakes. *Geophysical Research Letters* 46: 6425–6434.

Sweetkind D.S., Rytuba J.J., Langenheim V.E., Fleck R.J. (2011) Geology and geochemistry of volcanic centers within the eastern half of the Sonoma volcanic field, northern San Francisco Bay region, California. *Geosphere* 7: 629–657.

Wagner D.L., Saucedo G.J., Clahan K.B., Fleck R.J., Langenheim V.E., McLaughlin R.J., Sarna-Wojcicki A.M., Allen J.R., Deino A.L. (2011) Geology, geochronology, and paleogeography of the southern Sonoma volcanic field and adjacent areas, northern San Francisco Bay region, California. *Geosphere* 7: 658–683.

Watt J., Ponce D., Parsons T., Hart P. (2016) Missing link between the Hayward and Rodgers Creek faults. *Science Advances* 2: e1601441.

Woodrow D.L., Chin J.L., Wong F.L., Fregoso T., Jaffe B.E. (2017) Description of gravity cores from San Pablo Bay and Carquinez Strait, San Francisco Bay, California: U.S. Geological Survey Open-File Report 2017–1078.

3 Water

Water has also had a great effect on the Bay Area. Water is a powerful force, especially when it has millions of years to work. The water has come from precipitation, the run-off from rivers, and sea level rise and fall. These patterns have been in place essentially as long as the geological forces.

PRECIPITATION

California has typically experienced periods of more or less precipitation. In any year, two-thirds of the state's precipitation falls in Northern California. The cycles have covered decades of time. For the last 50 years, the state has been in a relative wet phase. The Western United States has experienced extreme fluctuations in rainfall. These extremes can easily be demonstrated by the 2012–2015 drought in California that was followed by winter storms that resulted in floods and mudslides in 2015–2016 (Barth et al., 2016). Barth et al. also noted that more than 80% of these severe weather events are associated with atmospheric rivers in Northern California (Fig. 3.1).

Most of the rain in the Bay Area is of average intensity, but periodically, California has been struck by atmospheric rivers. These are very large-scale precipitation events that involve 100–150 mm of rain in a 24-hour period seem to occur about every 2 years (Cordeira et al., 2019). However, those numbers can be much greater, up to 600 mm in some cases. In the winter of 1861–1862, 900 mm of rain fell in 30 days. The Sacramento and San Joaquin valleys were flooded, and more than 3 million acres were under water. Most of the annual rainfall and all of the extreme events can be attributed to these atmospheric rivers. These cause a number of deaths and a lot of property damage. These storms are caused by low-level jets along the front of warm sectors of winter cyclones in the eastern North Pacific (Dettinger et al., 2011). They carry very large amounts of water vapor in a long narrow path. They are often more than 2000 km long, but only a few hundred km wide, and they are concentrated in the lowest 2.5 km of the atmosphere. On satellite maps, they stretch from California back to Hawaii.

Landslides are an agent of major change in mountainous areas, even greater than erosion, and either shallow or deep seated, they are a particular hazard. Wild fires that clear off the vegetation leave hills more vulnerable without roots to add stability to the soil on slopes. That risk is enhanced if the ground has been saturated by previous rain. Cordeira et al. (2019) examined 142 years of landslide data (1871–2012) in Northern California and compared those to the incidence of atmospheric rivers. They found that most of the landslides occurred in the winter (particularly January and February) and that most occurred in coastal

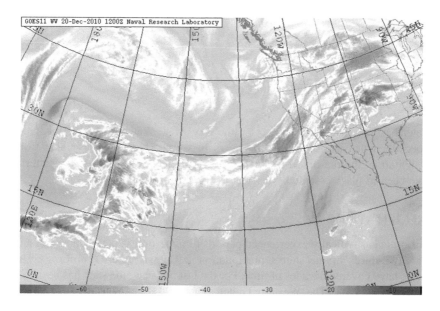

FIGURE 3.1 Atmospheric Rivers. In some cases, long narrow bands of water vapor form. They can stretch for thousands of miles, and when they reach California, they can release enormous amounts of rain in a very short time. The image is from the United States Naval Research Laboratory, Monterey, and is in the public domain.

counties that have steep slopes (e.g., Sonoma, Marin, San Mateo, and Santa Cruz counties). Interestingly, they also found that 82% of the 214 landslides that they studied occurred during an atmospheric river event.

As one might expect, drought tends to reduce the number of landslides. Bennett et al. (2016) examined this hypothesis during the great drought in Northern California in 2012–2015. They found that earth movements at less than 15 m did not consistently slow down in the drought. They speculated that individual storms might have a greater effect on these landslides than expected or groundwater conditions and vegetation might influence events. However, the number of earth movements of more than 15 m was definitely decreased. Landslides occur at different rates. The most dangerous move kilometers downslope at tens of meters per second. Slower, less dangerous ones—also called slow-moving landslides—move as slowly as millimeters per year and might stay active for decades (Handwerger et al., 2019). Handwerger et al. showed that even the slow-moving landslides in Northern California that had greatly speeded up in the record rain year of 2017 that broke the previous drought.

RUN-OFF

In California about 2 million years ago, the snow fell in the Sierra Nevada as it still does in winter. In the summer, it melted, and the water flowed downhill and drained into Monterey Bay by way of the Salinas River. However, some of the water remained in the Valley and slowly accumulated until the Valley was a vast lake.

For the last million years or so, the Bay has had its current structure. Geological forces work on a very long timescale. However, the geologic forces that have rocked California since its beginning were still active, and geology is not the only force at work in the Bay. As the plates moved, the surface land changed, and that happened in the great valley in the center of the state. The land of the Santa Lucia and Gabilan ranges near the southern end of the valley rose and cut off the water's path to the ocean. The water had nowhere to go, and so, it collected to form an enormous lake called Corcoran Lake. The lake was up to 1000 feet deep and stretched from modern-day Lake Shasta 500 miles to the Grapevine just north of Los Angeles (Fig. 3.2).

About 600,000 years ago, things changed dramatically. The size of the lake had changed over time, depending on climate conditions, and at one of its particularly high times, the water eroded the parts of the Diablo and Mayacamas ranges at the site of the present-day Carquinez Strait. The water broke through the hills with great force. Geologists believe that it might have happened suddenly so that the water ripped its way through to the Pacific creating deep channels in the Bay at Raccoon Strait between Angel Island and the main land and at the Golden Gate itself. The flood washed even more sediment into the Bay Area. By about 500,000 years ago, the lake was drained, and most of the central valley was dry. There remained a number of smaller lakes, and the Delta had formed behind the Carquinez Strait, where water would still sometimes be stopped. This allowed a great deal of sediment to be deposited in the Delta (Fig. 3.3).

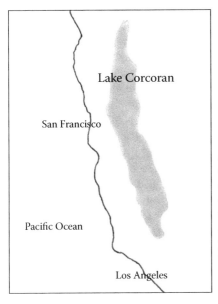

FIGURE 3.2 Lake Corcoran. Before about 650,000 years ago, much of Central California was covered by water. Lake Corcoran drained much of California and the water flowed to the Pacific Ocean through the Elkhorn Slough at the Monterey Bay. Geologic events cut off the flow to the Monterey Bay, and the water eventually broke through to cut the Carquinez Strait and a new path to the Ocean through what would become San Francisco Bay.

FIGURE 3.3 Carquinez Strait. As geologic forces caused upward and downward movements of the surface of the Bay Area, the path for water from Lake Corcoran to the Pacific changed. A breakthrough about 650,000 years ago allowed a massive torrent of water to tear through the Bay Area. It carved a path through the Carquinez Strait and formed a new outlet to the Pacific, the one that still operates today.

Now there was a path for water to flow from the Sierras along the California River through the Delta and the Bay Area and into the Pacific. So, the water from Corcoran Lake changed the landscape in the Bay Area and much more. Importantly, it changed the Bay into an estuary. From then on, the tides mixed the fresh water flowing into the Bay Area with the salt water of the Pacific Ocean.

For next several hundred thousand years, sea levels rose and fell as the Ice Ages cycled. About 15,000 years ago, the last cold period ended, and the ice began to melt. Sea levels rose some 300–400 feet.

About 15,000–20,000 years ago during the last Ice Age, the Bay looked quite different than now. In fact, for most of its existence, the Bay contained no water at all except for. It was a broad valley with only a couple of rivers flowing through it. The main river drained from the Carquinez Strait to the Pacific Ocean. That time was the last Ice Age, and much of the Earth's water was frozen in huge ice sheets that covered much of the Northern Hemisphere. Sea levels had dropped by about 300 feet, and the Bay had emptied. In fact, the ocean had retreated about 30 miles offshore. The present-day Farallon Islands were hills on a long plain. Alcatraz and Angel Island were also hills inside the dry Bay. Some idea of the terrain can be gained by the fact that just outside the Golden Gate, the water now is about 350 feet deep, but near Angel Island, it is only about 140 feet. Thus, the slope of the river in the Bay drops about 200 feet in a couple of miles. An enormous amount of water was flowing down that river, and some have speculated that there might have been a waterfall near the Golden Gate at one point.

This situation had happened several times before. As the Earth moved from Ice Age to Ice Age, sea levels rose and fell, and the Bay filled and emptied. The last Ice Age ended 8000–12,000 years ago. The glaciers melted, and the sea levels rose again. The Farallons became islands again. At some point, the water rose high enough to flow over the smaller rises that separated the Bay from the Pacific, and the Bay began to fill. The filling took some time, but eventually, the Bay assumed the water level we see today.

While that single event had a dramatic impact on the Bay Area, myriad smaller events have and still affect the Bay by bringing silt and mud into the Bay. In more recent times, humans have greatly influenced the run-off. Certainly, the highly destructive hydraulic mining activities during the Gold Rush contributed massive amounts of silt to the Bay. The mining in the Sierras destroyed whole mountains. The material from those mountains was washed down stream and into the Bay. Some streambeds were raised by as much as 6 m (Strange, 2008). That silt took more than 100 years to clear the Bay. As more people arrived in the Bay, rivers and streams that lead into the Bay have been dammed and channeled to control flooding. In other cases, run-off of raw sewage, mine tailings, industrial pollution, and other materials have seriously eroded the quality of Bay water. The rich Delta peat islands were very attractive to farmers, and they diked, drained, and channeled the Delta. By the 1930s, the Delta contained about 60 islands that were protected by 1700 km of levees.

The watershed of the Delta accounts for 40% of California. The demand for water grew with the burgeoning population, and large water projects attempted to contain the water for the twin use of people and agriculture. Most rivers were dammed. Water is even routed as far south as San Diego. Now just about 50% of the water that originally flowed through the Delta actually makes it to the Pacific.

SEA LEVEL RISE

Sea levels have risen and fallen many times as the Earth experienced Ice Ages in which massive amounts of water were frozen in the ice caps and glaciers that covered much of the Northern Hemisphere. The last of the Ice Ages occurred about 30,000–40,000 years ago. During that last Ice Age, the Bay was a dry plain with hills and a couple of rivers that ran out through the Golden Gate and across a plain some 50 km to finally reach the Pacific Ocean. As the Ice Age ended, about 10,000 years ago, sea levels rose, and water eventually flooded the Bay as we know it today. That pattern of a wet and dry Bay has repeated multiple times over the ages.

In 1977, Atwater et al. reported two sites of estuarine deposits that showed at least two times that the water in the Bay was substantially lower than it is today. They were the Sangamon and post-Wisconsin high stands of sea level (see Otvos, 2014, for a description of those times). When the water came back into the Bay 10,000 years ago, it entered at about 2 cm per year and spread about 30 m per year until about 8000 years ago when it slowed down to the present rate of about 0.1–0.2 cm per year.

DROUGHT

California and the Bay Area have experienced periods of wet and dry throughout its existence.

In recent years (2011–2014), the state has suffered through a multiyear drought with the lowest annual precipitation on record and the lowest in the last 1200 years (Griffin and Anchukaitis, 2014). These have resulted from a change in the usual storm track to the north that diverted rain to other areas.

Global warming has been a significant factor in the drought. Higher temperatures increase the rates of evaporation and intensify droughts. California depends on heavy snows in the Sierras in the winter that melt and run off during the summer.

Griffin and Anchukaitis (2014) used two methods to look at the history of drought in California. First, they looked at well-established tree ring chronologies. Second, they used their own measurements of recent tree rings in blue oaks (*Quercus douglasii*) from four sites. The blue oaks provide exceptionally strong moisture signals. They found that the lack of moisture is not unprecedented in California history, but they also suggest that the severity of the drought is intensified by the higher temperatures.

Tree rings are a good proxy for determining temperatures and precipitation. Thin rings indicate less water, and thick rings plenty of water. For example, scientists using this method found that, in 200 AD, a drought in the US Southwest lasted for 50 years.

Another method was used by Stine (1994). He examined tree stumps that were exposed when the level of Mono Lake decreased in the 1990s. Using radioactive carbon dating, he was able to determine when differences in water levels over time. He found two very long droughts. One began in the 9th century and lasted 200 years. A second began in the 13th century and lasted 150 years.

In the past, the people living in the Bay Area had little ability to deal with extended drought. Warmer temperatures and droughts occurred from A.D. 800 to 1350 as determined by analyses of pollen, tree rings, carbon isotopes, and fire scars on trees (Pilloud, 2006). The climate changes also stressed the Native American populations and may have caused social dislocations.

California already has extreme demands on the limited water resources available. A very real concern is that the water and agricultural infrastructure in the Bay Area and California have been built in what was an unusually wet period. If the west is normally much drier than it has been in recent years, then the area might be in for a difficult transition to deal with much less water.

REFERENCES

Atwater B.F., Hedel C.V., Helley E.J. (1977) Late Quaternary depositional history, Holocene sea-level changes, and vertical crustal movement, Southern San Francisco Bay, California. Geological Survey Professional Paper 1014.
Barth N.A., Villarini G., Nayak M.A., White K. (2016) Mixed populations and annual flood frequency estimates in the western United States: The role of atmospheric

rivers. *Water Resources Research* 53: 257–269. https://agupubs.onlinelibrary.wiley. com/journal/19447973.

Bennett G.L., Roering J.J., Mackey B.H., Handwerger A.L., Schmidt D.A., Guillod B.P. (2016) Historic drought puts the brakes on earthflows in Northern California. *Geophysical Research Letters* 43: 5725–5731. https://agupubs.onlinelibrary.wiley. com/doi/full/10.1002/2016GL068378.

Benson L., Kashgarian M., Rye R., Lund S., Paillet F., Smoot J., Kester C., Mensing S., Meko D., Landström S. (2002) Holocene multidecadal and multicentennial droughts affecting Northern California and Nevada. *Quaternary Science Reviews* 21: 659–682. https://digitalcommons.unl.edu/cgi/viewcontent.cgi?article=1374& context=usgsstaffpub.

Cordeira J.M., Stock J., Dettinger M.D., Young A.m., Kalansky J.F., Ralph F.M. (2019) A 142-year climatology of Northern California landslides and atmospheric rivers. *Bulletin of the American Meteorological Society* 100: 1499–1509. https://journals. ametsoc.org/bams/article/100/8/1499/344782.

Dettinger M.D., Ralph F.M., Das T., Neiman P.J., Cayan D.R. (2011) Atmospheric rivers, floods and the water resources of California. *Water* 3: 445–478.

Diffenbaugh N.S., Swain D.L., Touma D. (2015) Anthropogenic warming has increased drought risk in California. *Proceedings of the National Academy of Sciences of the United States of America* 112: 3931–3936.

Griffin D., Anchukaitis K.J. (2014) How unusual is the 2012–2014 California drought? *Geophysical Research Letters* 41: 2014GL062433.

Handwerger A.L., Fielding E.J., Huang M.-H., Bennett G.L., Liang C., Schulz W.H. (2019) Widespread initiation, reactivation, and acceleration of landslides in the northern California Coast Ranges due to extreme rainfall. *Journal of Geophysical Research* 124. https://doi.org/10.1029/2019JF005035.

Otvos E.G. (2014) The Last Interglacial Stage: Definitions and marine highstand, North America and Eurasia. *Quaternary International* 383: 158–173.

Pilloud M.A. (2006) The impact of the medieval climatic anomaly in prehistoric California: A case study from Canyon Oaks, CA-ALA-613/H. *Journal of California and Great Basin Anthropology* 26: 179–191.

Stine S. (1994) Extreme and persistent drought in California and Patagonia during mediaeval time. *Nature* 369: 546–549.

Strange C.J. (2008) Troubling waters. *BioScience* 58: 1008–1013.

4 Geomorphology of the Bay Area

INTRODUCTION

Geomorphology is the science of how the surface topography of the earth is a result primarily from the interactions of the environment upon the underlying geology (Hunt 1988). The resulting landscapes reflect not only the chemical and physical composition of the underlying rocks, but also the effect of water and weathering of those rocks, atmosphere and hydrosphere; the soils created from the rocks by the presence and flow of water; the microbes in the soils; the botanical diversities that are successful in the environment; and the combined influence of herbivores and predators within the ecological community, namely the biosphere. Finally, the effect of humans also may contribute significantly to the landscape and thereby may undermine some of the balancing forces that had been in effect for millennia.

This chapter describes a few examples of the geomorphology of regions within the San Francisco Bay Area that represent the results of different interactions between the environment, humans, and the earth's surface and which have created the landscapes we see today. We summarize the eight processes that affect geomorphology as they relate to the San Francisco Bay Area: aeolian, biological, fluvial, glacial, hillslope, igneous, tectonic, and marine. We will describe a number of localities in the Bay Area that demonstrate either a particular process or a combination of processes. These localities are of interest when viewed in the context of the Bay Area, including how the underlying rock influences the overlying vegetation and how past volcanism and plate tectonics result in the structures of the East Bay hills. We also summarize how precipitation and water flow, estimated over the past 2000 years, has affected the geomorphology of the Bay Area.

Geomorphological processes generally fall into three groups: (1) the production of rock and mineral fragments (regolith) by weathering and erosion, (2) the transport of that material, and (3) its eventual deposition and interaction with the underlying surface (Derbyshire et al., 1979). The processes are further categorized as follows.

AEOLIAN

Named for Aeolus, the Greek god of the wind, these are wind-generated geologic processes. Wind is characterized as the movement of air between atmospheric high-pressure and low-pressure systems, in particular at the interface between the atmosphere and the surface of the Earth. Depending upon the wind velocity, there will be a number of different effects that wind may have upon the underlying rocks, sand, silts, minerals, and the biosphere. Silts and sands are generally defined according to size, silts being smaller than 0.0625 mm down to 0.004 mm and sands being larger than 0.0625 mm up to 2 mm (ASTM, 1985). The effects can be transport, erosion, and deposition of rocks, sand, and minerals as well as the consequences upon vegetation and precipitation.

For wind velocities less than $10\,km.h^{-1}$ ($\sim 5\,m.s^{-1}$ at 1 m above the surface) sand particles remain in place (Sloss et al., 2012); however, at greater wind velocities ($\geq 5\,m.s^{-1}$ at 2.4 m above the surface) sand particles may be transported many tens or hundreds of meters; in addition, greater wind velocities can transport proportionally more sand particles than at lower velocities (Webb et al., 2016). Coarse sand particles, usually a feldspar or quartz, both high in silica, having more abrasive properties that an equivalent softer rock (for example, gypsum, apatite, or other minerals that make up much of sedimentary rocks), are particularly important in the process of erosion, whereby the force of the sand particle upon a rock or mineral surface is sufficient to break off fragments of the rock or mineral, either to be carried further by the wind or deposited in the vicinity, where the fragments may later be subjected to fluvial action (see below). Interestingly, vegetation cover can also influence the local wind velocity and erosion and deposition rates (Webb et al., 2016).

BIOLOGICAL

Biogeomorphological processes, perhaps unexpectedly to the layperson, have probably the most influence upon geomorphology of the land and the marine/freshwater environments than any of the other processes. First and foremost are those processes relating to the underlying geology, the subsequent weathering of those rocks by aeolian processes to create a fragmentary mineral layer, and the microbes and plants that use those minerals to grow and propagate: their waste products add to the biological and chemical detritus that, combined with the fragmentary minerals and chemical weathering, form the soil. The soil depends upon (1) the type of minerals released from the underlying rock; (2) the environmental conditions, including heat/cold, wet/dry conditions, and oxygen availability; and (3) the resulting ecosystem that is supported by that particular soil and environment. This may result in well-drained rolling hills having grass-covered hilltops and slopes, with deep valleys where streams rush to cut a course, and where woodlands only grow where the water is concentrated (see Fig. 4.1). Another example is to be found at fault line of the Pacific plate and the North American plate visible along the San Andreas Fault at the Crystal Springs reservoirs (Fig. 4.2). Here, there are several types of rock from the Franciscan Complex (FC) and the Great Valley Complex (GVC): the western side of the fault is composed mainly of FC early Cretaceous/Late Jurassic sedimentary and volcanic rocks, and Eocene sedimentary rocks; the eastern side comprises GVC Jurassic serpentine and FC Eocene/Paleocene/late

Cretaceous mélange, as well as FC early Cretaceous/late Jurassic sedimentary rocks (Graymer et al., 2006). The drainage patterns of the volcanic and sedimentary rocks on the west contrasts with that of the serpentine and mélange that border the San Andreas Fault resulting in a different vegetation.

Second, zoogeomorphology, whereby animals influence the form of the land, is an important process. Examples include beaver dams that modify the flow of water and sediment across the land, the action of burrowing animals, digging for tubers and roots, and the formation of nutrient-rich environments left by the roots and above-ground parts of trees and shrubs both before and after death, which can affect the way that the soil components are transferred from one layer to another, thereby providing new environmental conditions for organisms to take advantage of. Third, and perhaps less obvious, is that the combined ecosystem can influence the balance of atmospheric carbon dioxide, which ultimately can modulate the climate. One notable exception to the large influence of biological activity upon the geomorphology of a region is to be found in Antarctica, but of course this does not relate to the Bay Area.

FIGURE 4.1   Folding in the Bay Area. The mass of trees marks the bottom of Siesta Valley in the East Bay. It was not cut by erosion due to a stream at the bottom. It was formed by the folding in the surface that lifted the land thousands of feet on both sides. In a syncline, the rocks below the valley are "U" shaped.

FIGURE 4.2   View of the Crystal Springs Reservoir (San Mateo County) looking north. Photograph by Dick Lyon at Wikimedia.org and is licensed to share under Creative Commons. https://commons.wikimedia.org/wiki/File:Crystal_Springs_Reservoir_aerial_view_February_2018.JPG

FLUVIAL

Fluvial refers to any moving or stationary water body on the terrestrial landscape, such as creeks, streams, and rivers. Beginning from a spring or as meltwater from a glacier or snow, the water seeks the lower altitudes on its course its destination, either a lake or the sea. As it flows, it will erode the enveloping stream bank and its bed, forming a V-shaped valley; the amount of erosion will depend upon its velocity downhill. Greater velocities result in increased erosion, which translates into more geomorphological variation over time; fast-moving streams usually have a rather more straight path than those slower-moving waters but will deposit the sediment when suddenly slowed; alluvial fans on the sides of mountainous cliffs are an example of such flow and deposition. As the terrain becomes less steep, the water velocity slows, thus reducing the amount of bank and bed erosion and the path of the river may meander. Greater velocities upstream also result in the amount of sediment that may be transported and thus large volumes of sediment may be deposited as the river slows to a meandering phase. This in turn may build up the surface of the land through which the river flows, and which may result in changes in the ecosystems and landscape; the availability of new nutrients may encourage other species to proliferate in that environment; the slow pace and slower erosional rate may result in the formation of ox-bow lakes and fluvial terraces, a botanically diverse river system that provides additional biological niches for other animals and plants.

GLACIAL

Glaciers are not as significant a process upon the geomorphology of the Bay Area. The movement of glaciers down an old river valley slowly erodes the valley sides and the floor producing rock debris and creating a U-shaped valley; when the climate warms up, the glacier melts and retreats up the valley, leaving the rock debris on the surface, termed moraine. The San Francisco Bay Area was not subjected to direct effects of glaciation during the last Ice Age (115–15.2 thousand years ago (kya)) and the Younger Dryas cooling (13–11.1 kya); however, the Sierra Nevada range to the east was capped by an ice sheet for most of that time, having secondary effects upon both the climate and the geomorphology of the region (Kennett and Ingram, 1995). These secondary effects include meltwater run-off from the existing and melting glaciers, thus providing water flow from the central valley to the then-dry bed of the San Francisco Bay and the Carquinez Straits. Sediments bought down from the glaciers and the rivers contributed to the leveling of the bay basin floor.

HILLSLOPE

Soil, eroded minerals, sand, and rock will move down a slope as creep and accumulate at the base in the valley; the slope surface can be anything from essential vertical to almost flat and the angle will determine the rate of creep. The moment that creep begins depends upon (1) the rate of weathering of the rock; (2) the amount of water present within the soils; and (3) the composition of the underlying rock. Ongoing hillslope processes will change the topography of the hill's surface, resulting in a steadily increasing base height, and which will further retard the rate of creep towards the valley floor. In addition, animal activity as reported above may also affect the rate and onset of creep.

IGNEOUS

Igneous processes, defined as the result of volcanic activity (both eruptive and intrusive), may catastrophically alter the landscape, thereby resetting the paths of flowing water, migration routes of birds and animals, and delay the establishment of the botanical ecosystem for many years. In the Bay Area, excellent examples of igneous processes are found at the Sibley Volcanic Regional Preserve. Following the volcanic activity of 10 million years ago (mya), the basaltic lava was overlain by a succession of volcanic tuffs (ash) and fluvial sedimentary processes such as the Siesta Formation (Case, 1968).

Since then, the dates are not precise, the overlying sediments were eroded and the region was subjected to tectonic processes (see below), resulting in the creation of the Siesta Syncline, the great reverse fold that has exposed the basaltic and conglomerate rocks at the Preserve. For more details of the geomorphology of the Preserve, see Chapter 2.

TECTONIC

These processes are the result of the constant motion of the continental and oceanic plates, as described in Chapter 2. These are very complex in the Bay Area (D'Alessio et al., 2005). The resulting constant movement, which can be instantaneous or other hundreds of years, causes the relationship between adjacent ecosystems/geomorphology systems to be in constant flux, and can be observed as differences in vegetation and the animal life they support. A good visual example of such an environment is along the San Andreas fault at the Crystal Springs Reservoir, in particular where California Highway 92 links interstate I-280 with the coast at Half Moon Bay (see Fig. 4.2). Tectonic activity, such as earthquakes, may cause the upper portions of the crust to rise or collapse, generating more changes in the environment, fluvial flow, and weathering.

MARINE

Marine processes are those of the action of waves, marine currents, and seepage of fluids, including seawater mixed with decomposing organic material, through the seafloor. In the Bay Area, the long line of cliffs that run intermittently from north to south along the Pacific coast from the Marin headlands to Santa Cruz typify the ongoing onslaught upon the continuously mobile surface rock formations by the action of waves. The Bay Area is noted for the frequency of large amplitude waves that are generated thousands of miles away in storms from the eastern Pacific and which have begun to undermine many coastal communities whose properties were once hundreds of yards from the cliffs when they were built (Fig. 4.3).

SUMMARY

The landscapes surrounding the San Francisco Bay comprise mainly gently undulating hills and mountains interspersed by the lowlands of the bay shoreline and the floodplains of the San Joaquin and Sacramento Deltas and those of the Guadalupe River and Coyote Creek. These terrains suggest a history of steady rainfall on porous rock formations, thereby causing slow but persistent erosion of slowly uplifting crust. Empirical data derived from oxygen and carbon isotopic measurements of fossil bivalves (*Macoma nasuta*) contained in estuarine sediment were used to reconstruct a late Holocene record of salinity and stream flow in San Francisco Bay (Ingram et al., 1996).

FIGURE 4.3   Erosion at Pacifica. The cliff face is eroding under the pressure of waves striking the base of the cliff. Many homes and apartment complexes within a few hundred meters of the cliff have been or are slated for destruction before they fall into the ocean (Photograph courtesy of USGS).

The isotopic record suggested that between 2130 and 1670 calibrated years ago (ya cal. [baseline ya cal. = 1950]) (about 180 B.C.E. to 280 C.E.) there was substantially more inflow into the bay from the Central Valley than just prior to the mid-1800s, when diversion of the many rivers that fed into the Bay began in earnest (pre-diversion). This inflow probably came from at least two sources: local rainfall and snowmelt from the Sierra Nevada. However, between 1670 to 750 ya cal. (about 280–1200 C.E.) the heavy inflow was considerably reduced, possibly due to a major hydrologic event, such as a severe drought (Stine, 1994). Since 750 ya cal. (about 1200 C.E.) fluvial flow to the bay was about 1.5 times greater than that just prior to the pre-diversion. This fluvial flow has varied with a period of about 200 years and alternate wet and dry periods have occurred every 40 to 160 years (Ingram et al., 1996).

REFERENCES

ASTM (1985) Classification of Soils for Engineering Purposes: Annual Book of ASTM Standards, D 2487-83, *American Society for Testing and Materials* 4: 395–408.
Case J.E. (1968) Upper Cretaceous and Lower Tertiary Rocks; Berkeley and San Leandro Hills California. *Geological Survey Bulletin* 1251-J, USGS, US Dept.

D'Alessio M.A., Johanson I.A., Bürgmann R., Schmidt D., Murray M.H. (2005) Slicing up the San Francisco Bay Area: Block kinematics and fault slip rates from GPS-derived surface velocities. *Journal of Geophysical Research* 110: B06403.

Derbyshire E., Gregory K.J., Hails J.R. (1979) *Geomorphological Processes (Studies in Physical Geography)*. Oxford, United Kingdom, Butterworth-Heinemann, Elsevier.

Graymer R.W., Moring B.C., Saucedo G.J., Wentworth C.M., Brabb E.E., Knudsen K.L. (2006) *USGS/California Geological Survey*. Sacramento, California, US Geological Survey, https://pubs.usgs.gov/sim/2006/2918/sim2918_geolposter-stdres.pdf.

Hunt C.B. (1988) *Geology of the Henry Mountains, Utah, as Recorded in the Notebooks of G. K. Gilbert, 1875–76*, 167. Boulder, CO, Geological Society of America.

Ingram B.L., Ingle J.C., Conrad M.E. (1996) A 2000 yr record of Sacramento-San Joaquin river inflow to San Francisco Bay Estuary, California. *Geology* 24: 331–333.

Kennett J.P., Ingram B.L. (1995) Paleoclimatic Evolution of Santa Barbara Basin During the Last 20 k.y.: Marine Evidence from Hole 893A. In: *Proceedings of the Ocean Drilling Program, Scientific Results*, edited by Kennett J.P., Baldauf J.G., Lyle M. (eds), Vol. 146 (Pt 2).

Sloss C.R., Hesp P., Shepherd M. (2012) Coastal dunes: Aeolian transport. *Nature Education Knowledge* 3: 21.

Stine S. (1994) Extreme and persistent drought in California and Patagonia during Mediaeval time. *Nature* 369: 546–549.

Webb N.P., Galloza M.S., Zobek T.M., Herrick J.E. (2016) Threshold wind velocity dynamics as a driver of Aeolian sediment mass flux. *Aeolian Research* 20: 45–58.

5 Early Biology of the Bay

EVOLUTION OF THE BAY

In the previous chapters, we described the geology and geomorphology of the San Francisco Bay Area during the past 10 million years. This chapter is devoted to the flora and fauna that existed during those times up until the beginning of the Holocene Epoch (11,700 years ago) and that was partly influenced by the underlying rock makeup, the seascapes and landscapes that resulted from interactions between the underlying rock and minerals and the weathering elements. Those elements include wind, atmospheric vapor concentration, water and snow precipitation, followed by formation of streams and rivers, lakes, wetlands, uplands, and the San Francisco Bay Delta itself.

Since the Bay Area did not exist before the dinosaurs went extinct, no dinosaur fossils or any other terrestrial vertebrates are found there other than some marine reptiles, such as mosasaurs. In the Mesozoic Era (165 to 66 mya), California was actively forming, and much of it was covered with shallow inland seas that held a number of marine invertebrates and reptiles. Land plants included conifers, cycads, and ginkgoes. Radiolarians were very common and their dead shells resulted in deposits of chert that are very common in the Bay Area. In like manner, deposits of limestone found throughout the Bay Area are the shells of tiny marine organisms that were cemented together over time. During the Cretaceous, ammonites and oysters were present.

As we observed earlier, for much of its geological history the Bay Area was submerged beneath shallow coastal seas; from its creation during the late Jurassic Period (about 165 mya) until more recently (30 mya), the prehistoric Bay Area was largely affected by the subterranean motion of the Farallon Plate beneath that of the North American Plate, and which resulted in periodic volcanic activity as well as changes in sea level caused by tectonic upward and downward motion of the continental plates. Much of the sedimentary rock laid down within the Bay Area date from the mid- to late Mesozoic Era (165 to 66 mya) and the Paleocene Epoch (66 to 56 mya) and are embedded by mainly invertebrate fossils. These include mollusks such as belemnites (squid-like cephalopods), ammonites (nautilus-like cephalopods), bivalves (e.g., oysters, clams, and pelycopods, all having bilaterally symmetrical shells, that is, their shells are upon their sides), and gastropods (including freshwater snails); crustaceans, such as crabs (decapods; Bishop, 1988; Clites, 2020) and acorn barnacles (*Sessilia* spp.); brachiopods (lamp shells, filter-feeders having dorsal/ventral shells that are distantly related to mollusks and annelid worms; Cohen and Weydmann, 2005); corals; sea urchins (such as sand dollars, order Clypeasteroida); and plankton,

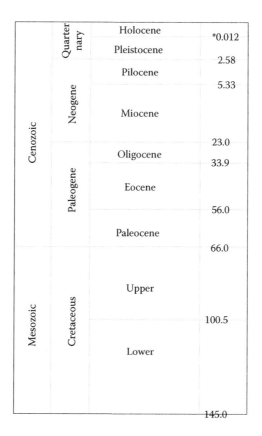

FIGURE 5.1   Geological Timeline. The asterisk indicates millions of years ago.

usually the larval forms of many invertebrate and some vertebrate species, and planktonic radiolaria (protozoa having mineralized skeletons) (Clites, 2020). There are also a smattering of vertebrate bones (e.g., whale vertebrae) in these marine deposits, but these are rare (Murray, 1974). What is not so readily apparent to the casual observer is the presence of microfossils, again usually larval forms of marine organisms; although invisible to the naked eye, they are of extreme importance to the energy industry geologist, for the presence (or absence) of a particular genus can indicate the co-localization of fossil fuels, such as oil and gas.

These marine organisms fluctuated both in numbers as well as across species, with the belemnites and ammonites becoming extinct at the end of the Cretaceous Period (66.5 mya), and increase in the numbers of crustacean species (Armstrong et al., 2009).

While the sedimentary deposits from the Mesozoic Era (which includes the Jurassic and Cretaceous Periods) are predominantly derived from marine or otherwise fluvial (rivers and lakes) conditions, the sedimentary rocks from the

Cenozoic Era (from the extinction of the dinosaurs 66 mya to the present day) comprise both aquatic and terrestrial species. The earliest sedimentary rocks found in the Bay Area from this Era are from the Miocene Epoch (from between 23 and 5.3 mya) in which are found barnacles and other shellfish, shallow-water inhabitants. By this time (about 5 mya) the Juan de Fuca Plate was almost entirely subducted beneath the North American Plate and the Pacific Plate had already accumulated marine sediments of its own and had become uplifted. Terrestrial sediments were thus forming on the eastern margins of the Pacific Plate and the western margins of the North American Plate and we find a plethora of plants and animal fossils from the Pliocene sediments.

MIOCENE AND PLIOCENE EPOCHS

At about 10 mya, the climate changed significantly. A number of plants from the age of the dinosaurs were the same as those existing today. For example, horsetails, ferns, and redwoods. Coast redwoods and giant sequoias were confined to narrow strips along the coast and in the Sierras, respectively.

The Bay Area was a very different place 9–10 mya. The Bay itself was not yet formed. Several small volcanoes were active in the East Bay hills. Mt. Diablo was still a broad plain and had not begun to be pushed up. Grasses and trees, including coast live oaks, elms, poplars, mahogany, and sumac, thrive in the mild climate. Some scientists have compared it to the modern-day Serengeti in Africa.

Several million years ago (9–10 million) in the Miocene Epoch, the basic form of the Bay Area was in place. However, there was no Bay. The Ice Age brought great masses of ice to North America and other places, and the sea levels dropped. By about a million years ago, rolling grasslands surrounded an enormous inland sea called Corcoran Lake covered what is now the Central Valley. Over time, geologic processes slowly transformed the land into the now familiar landscape.

Most of the environmental niches of today were filled, but with different animals. The Earth was dominated by mammals and birds. The largest were the elephant and mastodon relatives called *Gomphotherium*, and they lived alongside dozens of other large mammals, such as three-toed horse *Hipparion* the tiny pronghorn *Merycodus*, the long-necked camel *Aepycamelus* and peccaries. There are also smaller mammals, such as rabbits, beavers, ground squirrels, foxes. Birds and reptiles shared the space. The large number of herbivores supported a significant number of predators. The dog *Borophagus* probably killed and also ate carrion. *Nimravides* is a cat predator. The largest predator is the sabertooth *Barbourofelis*.

GREAT AMERICAN BIOTIC INTERCHANGE

The Great American Biotic Interchange (GABI) was an important paleozoogeographic event in which animals and plants were exchanged between the North and South American continents around the end of the Pliocene Epoch. Three million years ago, tectonic forces and subduction of the Cocos Plate under the Caribbean and North American Plates formed the Panama isthmus (O'Dea et al., 2016).

The presence of marsupials in North America may be seen to some as odd, since marsupials are usually associated with Australia. This is where an understanding of paleontology and geology brings to light the most likely explanation. During the early Cretaceous period, around 126 mya, the landmasses of North America, Europe, and Asia (composing the supercontinent of Laurasia) were in the process of breaking up; South America and Australia were still connected by Antarctica as part of the larger supercontinent Gondwana. At that time dinosaurs were the dominant large animal group and the ancestors of the marsupials, termed metatherians, were generally small and most likely nocturnal. Unfortunately, there is scant fossil evidence for these marsupial ancestors living on all three southern continents at that time; only monotremes and their ancestors have been found (Benson et al., 2013). It is most likely that the metatherians evolved in Asia, subsequently moved into North America by way of Europe, and then, at the end of the Cretaceous (66 mya), crossed by a landbridge to South America and then onwards expanding into Gondwana (Flynn and Wyss, 1998; Flynn et al., 2007). Plate tectonics then caused the landbridge to migrate northeast, where it became part of the Caribbean Archipelago, thereby cutting off any further connections between the continents of Laurasia and Gondwana (Kemp 2005; Boschman et al., 2014). There is, however, also evidence of non-placental therian fossils already in Gondwana during the mid-late Cretaceous (83.6–66 mya; Newham et al., 2014).

Thus the GABI enabled north-south movement of eutheria from North America to South America (cats, camels (e.g., vicuñas, the parent species of llamas), tapirs, and peccaries) and northward flow from South America of opossums and armadillos (O'Dea et al., 2016).

The Pleistocene (beginning about 2.6 mya) ushered in a long period of climatic instability, including successive cycles of glacial (ice ages) and interglacial (warmer periods) periods (Cohen et al., 2013). The reasons for this climatic instability are complex, but have been associated with wobble and periodic reversals of the Earth's magnetic field; fluctuations in the amount of energy from the sun, most likely due to eccentricities in the Earth's orbit around the sun; and tectonic/volcanic activity (Nordt et al., 2003; Foulger, 2010; Buis, 2020).

We will now present a brief description of the flora and fauna associated with the period from about 10 mya to the beginning of the Holocene Epoch, about 11,700 years ago.

FLORA

In the late Miocene and early Pliocene epochs (about 5–7 mya), a change occurred in the ratio of plants using the C3 and C4 photosynthetic pathways. For most plants on Earth, the first product of photosynthesis is a three-carbon product. In this process, because CO_2 enters through the stomata and the stomata are open, the plant can lose significant amounts of water. During a drought, this is a disadvantage. Plants in hot, dry areas have evolved C4 photosynthesis, which results in a four-carbon product. This process allows plants to carry on photosynthesis with the stomata closed. C4 plants include maize, sugarcane, and sorghum. The C4 process is also favored under conditions of low carbon dioxide levels. Cerling et al. (1997) examined the teeth of grazing animals during that period and determined that a significant change in the diets of those animals indicates a change in the atmospheric CO_2 levels that favored C4 photosynthesis.

California is one of the few regions in the world that features an exceptionally large number of unique plant species. That diversity stems from the late Tertiary period and is likely a function of the state's multiple environments with different altitudes and amounts of rainfall. However, Harrison et al. (2004) speculated that the diversity might also be a function of the presence of the mineral serpentine. Serpentine is an ultramafic rock that produces soils with less nutrients (N, P, K, Ca) than others and more toxins (Mg, Ni, Cr). Harrison et al. (2004) found that an association between the presence of serpentine and increased diversity exists, particularly in the Klamath and northern Coast Range. The age of exposure of the serpentine was an important factor. Anacker et al. (2010) further explored the influence of serpentine soils on plant diversity. They found that, once a plant lineage is specialized for its environment, it is less likely to further diversify than other plant lineages.

Plants were highly diverse by the time of the dinosaurs (Millar and Woolfenden, 2016). Conifers appeared about 200 mya. The pines were the last to develop, and they were common about 145 mya. Gymnosperms, including ferns and horsetails, expanded worldwide about 150 mya. Angiosperms, traced back to 110–150 mya, had expanded to become the most diverse group of plants on Earth. Angiosperms were in California at 100–120 mya, but their diversity can be pegged to about 55 mya. However, temperatures and humidity rose from 50 to 52 mya, and the angiosperms flowered (no pun intended). Species found in more tropical areas were common, including avocado, palm, viburnum, magnolia, jackfruit, and figs (Millar and Woolfenden, 2016). Pollen was found for pine, walnut, hickory, and sweetgum.

Adam et al. (1983) conducted an extensive examination of the plant and animal fossils from a site in the Bay Area near Saratoga. They found numerous mammal, bird, fish, mollusk, and ostracode fossils. Plants were represented by pollen and other remains. Interestingly, 50% of the pollen was from spruce trees (*Picea* spp.).

At about 33 mya, the tropical species disappeared and more temperature species again became dominant. Evergreen oaks, sycamore, cottonwood, willow, redbud, barberry, cherry, ironwood, manzanita, flannel bush, sumac, and grasses were found in Contra Costa County (Edwards, 2004).

Regarding plant species, at the boundary between the Miocene and the Pliocene (about 5.3 mya), alder, cherry, Christmas berry, chumico, coffee berry, dogwood, elm, flannel bush, Catalina ironwood, California lilac, magnolia, mountain mahogany, manzanita, live oak, poplar, bush poppy, swamp cypress, sumac, desert sweet, sycamore, tupelo, and willow all grew around the Bay Area (Murray, 1974).

FAUNA

As we mentioned earlier, animals that inhabited the Miocene environments are predominantly marine and riverine/estuary invertebrates, and these environments disappear from the fossil record as we enter the Pliocene Epoch (5.33 to 2.58 mya). This change in environment correlates with the uplift of the proto-Bay Area resulting from the relatively northward-moving Pacific Plate slipping along the San Andreas Fault up the western edge of the North American Plate, bringing much of what is now the northern and central California coastline up from present day southern California.

Although much of what is now the Bay Area was predominantly a marine environment during the Early to Late Miocene, fossils remains of the then-apex predators, beardogs (*Ysengrinia americanus*; Hunt, 2002, *Amphicyon ingens*, and *Cynelos jourdan*; Hunt and Yatkola, 2020) dating to between 15.8 and 14.0 mya have been found in the neighboring San Joaquin Valley in the California Central Valley. Beardogs were probably out-competed by hyenas at the Miocene/Pliocene transition (Wesley-Hunt, 2005).

Typical early Pliocene animal fossils include the traditional megafauna of North America. The term "megafauna" is considered to include megaherbivores (> 1000 kg) and megacarnivores (> 100 kg) (Malhi et al., 2016). In the early Pleistocene, these megaherbivores included camels (*Camelops*), ground sloths (*Megalonyx*), glyptodonts (*Glyptotherium*), toxodonts (*Toxodon*), oreodonts (Fam. Merycoidodontidea), and mastodons (*Mammut pacificus*), all of which evolved in the Americas. Large herbivores (45–999 kg) included primitive horses (such as the three-toed *Hipparion forcei* and *Merychippus californicus*), oreodonts, a medium-sized even-toed ungulate possibly related to camels, all now extinct (Spaulding et al., 2009); large carnivores (21.5–99 kg) included sabertooth big cats (such as *Smilodon*), American lions (*Panthera atrox*), canids (*Canis lepophagus* and *C. edwardii*, possibly the ancestors of the coyote and wolf, respectively), and hyenas (*Chasmaporthetes ossifragus*). Smaller mammals included foxes (*Vulpini*), giant beavers (*Castoroides nebrascensis*), primitive ground squirrels, mustelids (e.g., badgers, martens, and otters), peccaries (*Platygonus* and *Mylohyus*), raccoon-like animals (*Procyon* sp.), and birds and lizards of various species. The middle Pliocene fauna comprised beardogs, camels, flamingos (*Phoenicopterus copei* and *P. minutus*), ground sloths, mastodons, pronghorn antelope (*Antilocapra americana*), rhinoceroses (*Teleoceras*), cougar-like cats (*Puma pumoides*), and small rodents. The late Pliocene saw the appearance of many of North America's modern animal assemblage including bison (*Bison bison*), horses (*Equus* spp.), elk (*Cervus* spp.), and moose (*Alces americanus*) (Murray, 1974).

The Ice Ages began about 2 mya. By that time, many of the earlier mammals, including 80 genera, had disappeared. As is almost always the case, when one genus disappears, another takes its place. Those early mammals were replaced by mammoths, dire wolves, and sabertooth cats. Some sort of extinction event seems to have eliminated many of the living organisms, but not to the extent that it rises to a mass extinction event. The cause is unknown. Some speculate that climate change was involved. At about that time, the Earth cooled, the ice sheets at the polar caps expanded, glaciers expanded southward, rainfall was reduced, and the oceans cooled and sea levels fell. Other complications might have added to the stress on the mammals. Climate change also affected the vegetation. Grasslands gave way to a different environment more like our current one that features chaparral, multiple grasses, and many new species. Perhaps some combination of all of these was responsible for the loss of the genera. Some have speculated that the arrival of Native Americans might have contributed to the demise of many of these species. However, that hypothesis remains unproved at this point.

PLEISTOCENE AND HOLOCENE EPOCHS

The Pleistocene Epoch (from 2.6 million to 10,000 years ago) includes the climatic changes that brought about the periodic Ice Ages, as we discussed in Chapter 2 and above. It is here that the mammoths finally enter the North American ecosystem, having evolved around 5 mya in Africa from an ancestor that also gave rise to the Asian elephant (Krause et al., 2006; Lister and Bahn, 2007). The Columbian mammoths (*Mammuthus columbi*) in North America are descended from *Mammuthus trogontherii* that entered from Siberia over one million years ago (Lister and Sher, 2015). The woolly mammoth (*M. primigenius*) also evolved in Siberia and subsequently migrated across Beringia during an interglacial period possibly by around 400 thousand years ago (Lister and Sher, 2015). The Pleistocene Epoch also saw the appearance of the American lion (*Panthera atrox*), short-faced bear (*Arctodus simus*), modern horse (*Equus ferus/caballus*), steppe bison/bison (*Bison priscus/Bison bison*), reindeer/caribou (*Rangifer tarandus*), and musk ox (*Ovibos moschatus*) (Lorenzen et al., 2011).

Several times water has come and gone from the Bay Area. About 240,000 to 1.8 mya, it was a freshwater lake. During the peak of the last Ice Age, about 240,000-11,000 years ago, sea levels fell. The lake disappeared, and more land was uncovered. The Farallon Islands were hills on a plain 33-48 km west of the modern San Francisco. There still was no Bay. Rivers crossed it to flow out to sea through the Golden Gate. Radar studies of the present-day floor of the Bay reveals two major sub-surface structures, both probably related to the reduction in sea level during the last Ice Age. The first is a basin at what is now San Pablo Bay; this was most likely formed from a sudden mega-flood of melting ice from the western flanks of the Sierra Nevada. The second large structure appears to be a plunge pool that begins just east of the Golden Gate and reaches westward to the Marin Headlands, probably also resulted from the melt-water, but was sustained for some considerable time until sea levels rose over the lip of the Golden Gate waterfall; it was then followed by seawater inundation of the Bay itself and is known as the San Francisco Paleovalley (Cochrane et al., 2015; Dartnell, 2018).

During the late Pleistocene and the early Holocene, plants and animals evolved very little, but there were big changes in their populations and especially among the megafauna. All of these species and more died out and were not replaced by others as was the normal expectation. The reason is not clear. Some scientists suggest that climate change was responsible (Guthrie, 2003; Wroe and Field, 2006). However, others focus on the expansion of humans and their arrival in North America. Barnosky et al. (2016) weighed these two theories. First, they found that the average temperature changes and anomalous precipitation and the velocity of those changes 132,000 to 1000 years ago and found no correlation with the loss of mammals. From these results, they discount the effects of climate change on the disappearance of the megafauna in the early Holocene. Second, they evaluated the effects of humans on the animals. Their results suggest that early hominoids in Africa and those that left Africa (e.g., Neanderthals and Denisovans) co-evolved with the large mammals there. Once modern humans (*Homo sapiens*) began to leave Africa and spread into Australia, Northern Asia and Europe, and North and South America, they were the first humans to have contact with the mammals in those areas. Modern humans brought new tools and hunting techniques that easily and quickly overwhelmed the large mammals.

This story also played out in the Bay Area. Before about 12,000 years ago, large mammals wandered across the dry bed of what would become the Bay and wandered out along the plain to the hills that would later become the Farallon Islands, now 12 miles out to sea. The mammals included mammoths, mastodons, camels, horses, llamas, elk, tapirs, moose, and bison, along with large predators, such as the short-faced bear, sabertooth cat, wolves, and California lions.

LATE PLEISTOCENE EPOCH

About 100,000 years ago, the Late Pleistocene Epoch began. At that time, the geology of the Bay Area was well formed, but the Bay was a dry plain covered in grass with tree-lined rivers. Outside the Golden Gate, the plain extended many miles past the Farallon Hills (later islands). Large herds of mammoth, mastodons, camels, horses, bison, llamas, elk, and tapirs wandered about (Parkman, 2006). Over them all soared the condor (*Terratornis merriam*) with its wingspan of over 3 meters. The herds likely moved from inland in the winter to the shore in the summer to escape the heat and eat the vegetation kept lush by the fog. These would have been fairly short migrations of less than 60 km.

Among the large mammals were the Columbian mammoth (*Mammuthus columbi*), which weighed more than 5 tonnes and stood 4 meters at the shoulders. Long-horned bison (*Bison latifrons*) and ancient bison (*Bison antiquus*) both grazed on grasses. Both were larger than modern bison. The long-horned bison lived alone or in small groups, but the ancient bison were herd animals. American mastodon (*Mammut americanum*) was distantly related to elephants. It was shorter and stockier than the mammoth. The western horse (*Equus occidentalis*) was small and stood only about 1.5 meters at the shoulders. The giant horse (*Equus pacificus*) was larger. Interestingly, the large-headed llama (*Hemiauchenia macrocephala*) was a grazer and far from the Andes that are normally associated with llamas. Jefferson's ground sloth (*Megalonyx jeffersoni californicus*) was as large as an ox. Camels (*Camelops hesternus*) were herd animals of about the same size as modern camels.

Well-fed predators were there too, including the short-faced bear, sabertooth cat, dire wolf, and California lion. Dire wolf (*Canis dirus*) was about the size of a modern-day timber wolf and probably hunted in packs as do modern wolves. The large number of dire wolves found in the La Brea Tar Pits suggests that they ate carrion as well as hunted. Coyote (*Canis orcutti*) is now extinct. A large number of them were also found in the La Brea Tar Pits, and they were probably significant predators. They were larger than their modern counterparts, but could not make adjust after most of the larger mammals were eliminated. Giant short-faced bear (*Arctodus simus*) was the largest carnivore that ever lived in the New World. An adult weighed about 900 kg and was a third larger than a grizzly bear. In addition, its longer legs suggest that it was faster even than modern bears. The cat family was well represented. The Scimitar cat (*Homotherium serum*) was closely related to the sabertooth cat (*Smilodon californicus*) and hunted large animals. In addition to many of the modern cat forms, the sabertooth cats were major predators. They were slower than other cats and probably preyed on larger animals. The American lion (*Panthera atrox*) was larger than modern-day cats in Eurasia and may have hunted in prides as modern lions. The jaguar (*Panthera onca*), American cheetah (*Acinonyx trumani*), cougar (*Felis daggetti*), and lynx (*Lynx rufa fischeri*) rounded out this group.

The Quaternary Glaciation began about 2.58 mya and is ongoing (Marshall, 2010). From then to the present, the Earth has experienced several glacial periods. In that time, there have been several cycles of expansion and retreat of ice. In the colder parts of the cycle, ice sheets and glaciers covered a significant portion of the Northern Hemisphere. Temperatures and sea levels fell. Between those, temperatures and sea levels rose. For much of the Quaternary, the ice cycle repeated about every 41,000 years, but about a million years ago, the cycle changed to every 100,000. The reasons are not quite clear. In fact, the cause of the cycles themselves are not know, but seem to be related to the tilt of the Earth.

The last of these Ice Ages began 33,000 years ago and started to recede 19,000 years ago. During that time, sea levels were much lower than before, and a land bridge was revealed between Siberia and Alaska. Native Americans crossed that bridge to become the first humans in the America. They made it to the Bay Area about 15,000 years ago.

THE GREAT EXTINCTION

As the last Ice Age began to come to an end about 15,000 years ago, almost all the megafauna of North America became extinct. The causes of these extinctions have been debated amongst scientists for many decades and have revolved from (1) adaptations (or lack of) to climate change; (2) a reduction in habitat resulting in reduced food supply for these huge mammals; (3) the appearance of man, who may have hunted them to extinction; or (4) a combination of all these factors (Martin, 1984; Alroy, 2001; Barnosky et al., 2004; Stuart et al., 2004; Koch and Barnosky, 2006; Sandom et al., 2014; Surovell et al., 2016). A study that modeled population dynamics of six megafauna species over the past 50,000 years suggested that climate change and population fragmentation, and for certain species, human engagement, played the major role in their extinction in North America (Lorenzen et al., 2011). The remaining medium-to-large herbivores, such as bison, elk, moose, reindeer, and pronghorn antelope, survived this period of change, for reasons as yet unknown; this may be a combination of surviving in large herds, adapting to the relatively warmer climates and the multiplicity of food available in the woods and grasslands that came to dominate the landscape following the Ice Age. Musk ox survived in the Canadian and Alaskan Arctic (Lorenzen et al., 2011). In addition, populations of moose (*Alces*) and elk (*Cervus* spp.) increased towards the end of the Late Quaternary Epoch (late Pleistocene) that may have led to increased competition for both woodland and grassland resources (Guthrie, 2006). Of note, both reindeer (in Eurasia) and caribou (North America) have been hunted and/or herded by Arctic peoples for at least the past 10–15 thousand years (Kurtén, 1968), thus perhaps inadvertently preserving pan-regional breeding populations.

Many other cases of animals going extinct or nearly extinct are well known. The Pacific gray whale (*Eschrichtius robustus*) was hunted nearly to extinction. Between 1846 and 1874, as many as 8000 gray whales were killed by whalers. Other hunts continued until 1936 when gray whales became protected by the United States. Today they have rebounded. The Columbian mammoth (*Mammuthus columbi*) lived in the Bay Area during the Late Pleistocene. These relatives of modern elephants stood about 4 meters tall weighed 10.4 tonnes with tusks almost 5 meters long. They migrated into North America 2 mya but became extinct about 11,000 years ago. Other mammals included 2-ton bison, Western camels, giant horses, tapirs, and giant ground sloths. Perhaps 800 American lions lived in the Bay Area. They weighed 340 kg were much larger than modern African lions. The other major predator was the short-faced bear. They were about 4 meters long and weighed over 900 kg and could run at 65 kmph for about 2 km. Other predators included dire wolves, sabertoothed tigers, grizzly bears, and giant condors.

California originally had two types of bears. The black bear (*Ursus americanus*) was found as far south as Sonoma County. The California grizzly bear (*U. arctos californicus*) was found throughout the state. Thus, any reference to bears in early writings almost surely refers to grizzlies. The California grizzlies were larger than the grizzlies that lived in the Rockies (900 kg as compared to just under 700 kg). By the mid-19th century, cattle had become a major industry, and the grizzlies were a problem. Farmers began killing them. Within 75 years of the discovery of gold, California grizzlies were extinct.

REFERENCES

Adam D.P., McLaughlin R.J., Sorg D.H., Adams D.B., Forester R.M., Repenning C.A. (1983) An Animal and Plant-Fossil Assemblage from the Santa Clara Formation (Pliocene and Pleistocene), Saratoga, California. In: *Cenozoic Nonmarine Deposits of California and Arizona*, edited by Raymond V. Ingersoll, Michael O. Woodbourne (eds), pp. 105–110. San Francisco, CA: Society of Economic Paleontologists and Mineralogists, Pacific Section.

Alroy J.A. (2001) Multispecies overkill simulation of the end-Pleistocene megafaunal mass extinction. *Science* 292: 1893–1896.

Anacker B.L., Whittall J.B., Goldberg E.E., Harrison S.P. (2010) Origins and consequences of serpentine endemism in the California flora. *Evolution* 65: 365–376.

Armstrong A., Nyborg T., Bishop G., Osso A., Vega F.J. (2009) Decapod crustaceans from the Paleocene of central Taxas, USA. *Revista Mexicana de Ciencias Geologicas* 26(3): 745–763.

Barnosky A.D., Koch P.L., Feranec R.S., Wing S.L., Shabel A.B. (2004) Assessing the causes of late Pleistocene extinctions on the continents. *Science* 306: 70–75.

Barnosky A.D., Lindsey E.L., Willavicencio N.A., Bostelmann E., Hadly E.A., Wanket J., Marshall C.R. (2016) Variable impact of late-Quaternary megafaunal extinction in causing ecological state shifts in North and South America. *Proceedings of the National Academy of Sciences USA* 113: 856–861.

Benson R.B.J., Mannion P.D., Butler R.J., Upchurch P., Goswami A., Evans S.E. (2013) Cretaceous tetrapod fossil record sampling and faunal turnover: Implications for biogeography and the rise of modern clades. *Paleogeography, Paleoclimatology, Paleoecology* 372: 88–107.

Bishop G.A. (1988) Two crabs, *Xandaros sternbergi* (Rathbun 1926) n. gen., and *Icriocarcinus xestos* n. gen., n. sp., from the Late Cretaceous of San Diego County, California, USA, and Baja California Norte, Mexico. *Transactions of the San Diego Society of Natural History* 21: 245–257.

Boschman L.M., van Hinsbergen D.J.J., Torsvik Trond H., Spakman W., Pindell J.L. (2014) Kinematic reconstruction of the Caribbean region since the Early Jurassic. *Earth-Science Reviews* 138: 102–136.

Buis A. (2020) Milankovitsch (Orbital) Cycles and their Role in Earth's Climate. *NASA*, https://climate.nasa.gov/news/2948/milankovitch-orbital-cycles-and-their-role-in-earths-climate/, accessed September 28, 2020.

Cerling T.E., Harris J.M., MacFadden B.J., Leakey M.G., Quade J., Eisenmann V., Ehleringer J.R. (1997) Global vegetation change through the Miocene/Pliocene boundary. *Nature* 389: 153–158.

Clites E. (2020) Fossils in our Parklands. *University of California Museum of Paleontology*, https://ucmp.berkeley.edu/science/parks/golden_gate.php, accessed September 12, 2020.

Cochrane G.R., Cochran S.A., Cochrane G.R., Johnson S.Y., Dartnell P., Greene H.G., Erdey, Mercedes D., Golden N.E., Hartwell S.R., Endris C.A., Manson M.W., Sliter R.W., Kvitek R.G., Watt J.T., Ross S.L., Bruns T.R. (2015) https://pubs.usgs.gov/of/2015/1068/pdf/ofr2015-1068_sheet1.pdf, accessed August 15, 2020.

Cohen B.L., Weydmann A. (2005) Molecular evidence that phoronids are a subtaxon of brachiopods (Brachiopoda: Phoronata) and that genetic divergence of metazoan phyla began long before the early Cambrian. *Organisms Diversity & Evolution* 5: 253–273.

Cohen K.M., Finney S.C., Gibbard P.L., Fan J.-X. (2013) International Chronostratigraphic Chart 2013. *ICS, stratigraphy.org*, accessed January 7, 2019.

Dartnell P. (2018) US Geological Survey. *Pacific Coastal and Marine Science Center*, https://www.usgs.gov/centers/pcmsc/science/california-seafloor-mapping-program?qt-science_center_objects=0#qt-science_center_objects, accessed August 15, 2020.

Edwards S.W. (2004) Paleobotany of California. *The Four Seasons* 4: 3–75.

Flynn J.J., Wyss A.R. (1998) Recent advances in South American mammalian paleontology. *Trends in Ecology and Evolution* **13**(11): 449–454. doi:10.1016/S0169-5347(98)01457-8. PMID 21238387.

Flynn J.J., Wyss A.R., Charrier R. (2007) South America's missing mammals. *Scientific American*. **296**(May): 68–75.

Foulger G.R. (2010) *Plates vs. Plumes: A Geological Controversy*. Hoboken, New Jersey, Wiley-Blackwell.

Guthrie R.D. (2003) Rapid body size decline in Alaskan Pleistocene horses before extinction. *Nature* 426: 169–171.

Guthrie R.D. (2006) New carbon dates link climatic change with human colonization and Pleistocene extinctions. *Nature* 441: 207–209.

Harrison S., Safford H., Wakabayashi J. (2004) Does the age of exposure of serpentine explain variation in endemic plant diversity in California? *International Geology Review* 46: 235–242.

Hunt R.M. (2002) Intercontinental migration of Neogene Amphicyonids (Mammalia, Carnivora): Appearance of the Eurasian beardog Ysengrinia in North America. *American Museum Novitates* 3384: 1–53.

Hunt R.M. Jr., Yatkola D.A. (2020) A new species of the amphicyonid carnivore *Cynelos* Jourdan, 1862 from the early Miocene of North America. (In: de Bonis L., Werdelin L. (eds), *Memorial to Stéphane Peigné: Carnivores (Hyaenodonta and Carnivora) of the Cenozoic*) *Geodiversitas* 42(5): 57–67.

Kemp T.S. (2005). *The Origin and Evolution of Mammals*. Oxford, Oxford University Press.

Koch P.L., Barnosky A.D. (2006) Late Quaternary extinctions: state of the debate. *Annual Review of Ecology, Evolution, and Systematics* 37: 215–250.

Krause J., Dear P.H., Pollack J.L., Slatkin M., Spriggs H., Barnes I., Lister A.M., Ebersberger I., Pääbo S., Hofreiter M. (2006) Multiplex amplification of the mammoth mitochondrial genome and the evolution of Elephantidae. *Nature* 439: 724–727.

Kurtén Björn (1968) *Pleistocene Mammals of Europe*. Piscataway, NJ, Transaction Publishers, pp. 170–177

Lister A., Bahn P. (2007) *Mammoths: Giants of the Ice Age*. London, United Kingdom, Frances Lincoln LTD. p. 23.

Lister A.M., Sher A.V. (2015) Evolution and dispersal of mammoths across the Northern Hemisphere. *Science* 350: 805–809.

Lorenzen E.D., Nogués-Bravo D., Orlando L., Weinstock J., Binladen J., Marske K.A., Ugan A., Borreegaard M.K., Gilbert M.T.P., Nielsen R., Ho S.Y.W., Goebel T., Graf K.E., Byers D., Stenderup J.T., Rasmussen M., Campos P.F., Leonard J.A., Koepfli K.-P., Froese D., Zazula G., Stafford Jr. T.W., Aaris-Sørensen K., Batra P., Haywood A.M., Singarayer J.S., Valdes P.J., Boeskorov G., Burns J.A., Davydov S.P., Haile J., Jenkins D.L., Kosintsev P., Kuznetsova T., Lai X., Martin L.D., McConald H.G., Mol D., Meldgaard M., Munch K., Stephan E., Sablin M., Sommer R.S., Sipko T., Scott E., Suchard M.A., Tikhonov A., Willerslev R., Wayne R.I., Cooper A., Hofreiter M., Sher A., Shapiro B., Rahbek C., Willerslev E. (2011) Species-specific responses of Late Quaternary megafauna to climate and humans. *Nature* 479: 359–364.

Malhi Y., Doughty C.E., Galetti M., Smith F.A., Svenning J.-C., Terborgh J.W. (2016) Megafauna and ecosystem function from the Pleistocene to the Anthropocene. *Proceedings of the National Academy of Sciences USA* 113: 838–846.

Marshall M. (2010) The history of ice on Earth. *The New Scientist,* https://www.newscientist.com/article/dn18949-the-history-of-ice-on-earth/, accessed September 10, 2020.

Martin P.S. (1984) Prehistoric Overkill: The Global Model. In: *Quaternary Extinctions: A Prehistoric Revolution,* edited by Martin P.S., Klein R.G. (eds), pp. 364–403. Tucson, AZ: University of Arizona Press.

Millar C.I., Woolfenden W.B. (2016) Ecosystems Past: Vegetation Prehistory. In: *Ecosystems of California,* edited by Harold Mooney, Erika Zavaleta (eds), pp. 131–154. Berkeley, CA: University of California Press.

Murray M. (1974) *Hunting for Fossils: A Guide to Finding and Collecting Fossils in All 50 States.* New York, New York, Collier Books. 348, pp. 97–102.

Newham E., Benson R., Upchurch P., Goswami A. (2014) Mesozoic mammaliaform diversity: The effect of sampling corrections on reconstructions of evolutionary dynamics. *Paleogeography, Paleoclimatology. Paleoecology* 412: 32–44.

Nordt L., Atchley S., Dworkin S. (2003) Terrestrial evidence for two greenhouse events in the latest cretaceous. *GSA Today* 13(12): 4–9. doi:10.1130/1052-5173(2003) 013<4:TEFTGE>2.0.CO;2.

Nyborg T.G., Vega F.J., Filkorn H.F. (2003) New Late Cretaceous and Early Cenozoic decapod crustaceans from California, USA: implications for the origination of taxa in the eastern North Pacific. *Contributions to Zoology* 72: 165–168.

O'Dea A., Lessios H.A., Coates A.G., Eytan R.I., Restrepo-Moreno S.A., Cione A.L., Collins L.S., de Queiroz A., Farris D.W., Norris R.D., Stallard R.F., Woodburne M.O., Aguilera O., Aubry M.P., Berggren W.A., Budd A.F., Cozzuol M.A., Coppard S.E., Duque-Caro H., Finnegan S., Gasparini G.M., Grossman E.L., Johnson K.G., Keigwin L.D., Knowlton N., Leigh E.G., Leonard-Pingel J.S., Marko P.B., Pyenson N.D., Rachello-Dolmen P.G., Soibelzon E., Soibelzon L., Todd J.A., Vermeij G.J., Jackson J.B. (2016) Formation of the Isthmus of Panama. *Science Advances* 2(8):e1600883.

Parkman E.B. (2006) The California Serengeti: Two Hypotheses Regarding the Pleistocene Paleoecology of the San Francisco Bay Area. http://www.parks.ca.gov/pages/22491/files/the_california_serengetti_pleistocene_paleoecology_of_san_franc isco_bay.pdf, accessed September 10, 2020.

Sandom C., Faurby S., Sandel B., Svenning J.-C. (2014) Global late Quaternary mega-fauna extinctions linked to humans, not climate change. *Proceedings: Biological Sciences* 281: 1–9.

Spaulding M., O'Leary M.A., Gatesy J. (2009) Relationships of Cetacea (Artiodactyla) among mammals: increased taxon sampling alters interpretations of key fossils and character evolution. *PLoS One* 4(9): e7062.

Stuart A.J., Kosintsev P.A., Higham T.F.G., Lister A.M. (2004) Pleistocene to Holocene extinction dynamics in giant deer and woolly mammoth. *Nature* 431: 684–689.

Surovell T.A., Pelton S.R., Anderson-Sprecher R., Myers A.D. (2016) Test of Martin's overkill hypothesis using radiocarbon dates on extinct megafauna. *Proceedings of the National Academy of Sciences USA* 113: 886–891.

Wesley-Hunt G.D. (2005) The morphological diversification of carnivores in North America. *Paleobiology* 31: 35–55.

Wroe S., Field J. (2006) A review of the evidence for a human role in the extinction of Australian megafauna and an alternative interpretation. *Quaternary Science Reviews* 25: 2692–2703.

6 Humans Arrive

NATIVE AMERICANS

The first settlers in the Bay Area were the Native Americans. The last Ice Age had sequestered massive amounts of water in the polar ice cap and in glaciers that covered much of the North America. With so much water frozen away, sea levels dropped dramatically and exposed new dry land that had previously been covered with water. One aspect of the lower sea levels was the appearance of a land bridge between Siberia and Alaska. Humans took advantage of that land bridge to cross into North America. The dates for the migration are the subject of continuing research. However, they seem to have migrated into the New World in two or more waves, beginning around 16,500 years ago, and rapidly spread out across North and South America. They arrived in the Bay Area over 10,000 years ago. Ultimately, about 30 tribes lived in California. By the time of contact with Europeans, the number of Native Americans in California was estimated at 300,000.

When the first Native Americans arrived, sea levels were much lower than they are today. The Bay was still a wide valley with rivers running through it, and a plain spread out 30 some miles beyond the Golden Gate and past the Farallon Hills as they were then (Fig. 6.1). The climate was temperate, and food was readily available. These first inhabitants were hunter-gatherers, and they did not till the soil. Thus, their effects on the environment were minimal, especially compared to those of the Europeans who arrived later. However, they were not without consequences.

The diet of the Native Americans in the late Holocene is unclear (Bartelink, 2009). Fish and other seafood were readily available, and these were a major part of the diet of the Coastal tribes. Those tribes constructed more than 400 shell mounds in the Bay Area and used the mussel shells to make beads for trade. The availability of large game declined, and plant foods became more important. The reason is thought to have been overhunting, but changes in climate might have been involved. The use of seafood also diminished (Bartelink, 2009).

The Native Americans used forest gardening to secure food. Others more inland conducted agriculture. At the time, much of the Bay Area hills consisted of scrub lands. These woody plants have little food value. The patterns of grass and shrubs cause some researchers to speculate that the coastal tribes were more concentrated and that they used fire to clear land (Keeley, 2002). Lightning is not common in California, and so, few fires begin from lightning strikes. The researchers believe that the Native Americans intentionally set fires to burn off the underbrush. This practice prevented larger fires, renewed the land, and encouraged oaks to grow.

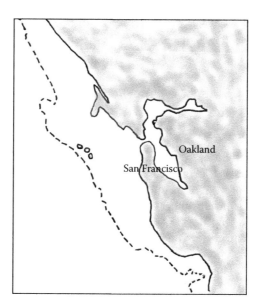

FIGURE 6.1   Changing California Coastline. The coastline is a function of sea level. At the height of the last Ice Age (20,000 years ago), sea levels were about 90 m below what they are today. The Bay was dry except for a few rivers, and the coast was about 50 km (dotted line) west of where it is today. The Farallon Islands (represented by three small dots) were hills in a long plain that stretched out into what is today the Pacific Ocean. As the Ice Age ended, sea levels rose again and covered the plain, leaving the Farallon as islands. Eventually, the water rose up enough to pour over the Golden Gate and slowly fill the Bay to its level today.

They used the acorns for food. With regular burning, the shrubs can be replaced by more productive grasslands and black oaks. The fires they set also encouraged the growth of berries and other foods and controlled insects. The thick bark of the oaks protects them from fire, but intense fires can kill the trees (Long et al., 2017). However, on balance, the oaks fare better in fires than conifers, and they also bounce back more effectively after the fire. The use of fire likely had a significant effect on the environment by changing the chaparral and other shrublands to grasslands, at least in those areas near their villages. That change was desirable to increase game and to protect against predators and enemies. In the southern part of the Bay Area, the Native Americans began the process of changing the oak sa-vannahs from sage scrub to grasslands. The researchers believe that the Native Americans changed the nature of a considerable portion of the Bay Area, and that European settlers later continued this practice.

Black oaks (*Quercus kelloggii*) were a major food source for the Native Americans in the Bay Area, and they stored thousands of pounds of acorns (Anderson, 2007). They also liked tanoak (*Notholithocarpus densiflorus*) and others. Black oaks are common from Oregon to Southern California and provide homes for many animals, including mule deer (*Odocoileus hemionus*), acorn woodpeckers (*Melanerpes formicivorus*), pileated woodpeckers (*Dryocopus pileatus*), mountain quail (*Oreortyx pictus*), and band-tailed pigeons (*Patagioenas fasciata*), that the Native Americans needed (Long et al., 2017). In addition to acorns, the Native Americans gathered various seeds of grasses and forbs, legumes, mushrooms, and more. They did not conduct agriculture per se, but they did manage and cultivate wild areas.

One of the main tribes in the Bay Area was the Ohlone. They lived in more than 50 villages that averaged about 200 residents. They were mostly hunter-gatherers and also planted some crops. Acorns were a staple for the Ohlone, but they also ate nuts, grass seeds, and berries, as well as fish and seafood. They hunted elk, pronghorn, deer, ducks, geese, quail, and more. They lived in houses of woven mats of tules. In other cases, they used redwood bark over a wooden framework. Men generally went about naked in summer, and in winter, they wore animal skin capes. Women wore aprons or skirts of deerskin or tules.

For much of the last 4000 or so years, the climate in the Bay Area has been mild (Schick, 1994). Winter have been wet and green, and summers have been warm and dry. The amount of rain varies among the many microenvironments. However, there is typically more rain on the west side of the Coast Range and less in the Valley. It rarely ever snows other than on the peaks of the tallest surrounding mountains, and summer temperatures are usually less than 100 °F.

Oaks do well in these different environments, and their acorns are an important component of the Ohlone diet. The Coast Live Oak (*Quercus agrifolia*) prefers the coastal plains. Blue Oaks (*Quercus douglasii* Hook) like the hot interior valleys. Black Oaks (*Quercus kelloggii*) like higher elevations. The Interior Live Oak (*Quercus wislizeni*) seems to prefer riverbanks. The most important of the oaks to the Ohlone were the Coast Live, Black, and the Valley Oaks (*Quercus lobata*). The Valley Oaks are the largest of the oaks, and they produce the largest acorns.

Acorns are rich in carbohydrates and fats, and they can feed a population year-round. But it is not only humans who harvest and store acorns. A number of animals do the same, including bears, pigs, birds, squirrels, and rats. With the acorns, berries, and other nuts, an entire animal community is sustained.

In addition to the oaks, many other plants grow in the Bay Area. The oaks are linked to grasses and chaparral, which provide fodder for animals, such as antelope, deer, and elk. The other plant communities include grasslands, woodlands, evergreens, marshlands, and chaparral. Most of the natural habitats have been since destroyed by human activities. The Ohlones used many of the natural products. In the grasslands, they used willow, rush, tule, and other grasses to make baskets for collecting acorns, tule boats, and mats. They burned the chaparral to control the growth of shrubs and grasses and to allow new oaks to germinate and grow.

SHELL MOUNDS

The Native Americans lived in the Bay Area thousands of years without significantly transforming the land or Bay. Shell mounds are one of the most distinctive features left by the early Native Americans in the Bay Area. The mounds were constructed over thousands of years and are impressive in size and number. They range in size from 9 to 183 m in diameter and are up to 9 m tall. Finally, some extend below ground too. Some of the largest are found in Emeryville, Ellis Landing, West Berkeley, and the Stege Mound Complex in Richmond. More than 400 mounds have been identified, but many others were undoubtedly lost over the years to erosion and human activities. They are composed of large amounts (e.g., many cubic meters) of sediment, shells (e.g., clams, mussels, and oysters), rocks, and ash. Mounds in the North Bay and Sacramento Delta tend to use fewer shells and more mud and sand. The mounds also contained human and animal remains and a large number of artifacts.

The purpose of the mounds has been and continues to be the subject of considerable debate among archeologists. Early researchers viewed them as a simple midden or trash pile where hunter-gathers disposed of shells. More recent researchers see them as more intentional. They might have been used for religious purposes or even as living quarters.

EXTINCTION OF LARGE MAMMALS

The end of the Pleistocene era saw the extinction of 35 genera of large mammals, such as mammoths, mastodons, giant beavers, and others. These events seem to coincide with the arrival of the first Native Americans in North America. The "overkill hypothesis" (Martin, 1967) posits that this association shows that Native Americans hunted those mammals to extinction. Becerra-Valdivia and Higham (2020) analyzed the timing of 42 archaeological sites and integrated that information with genetic and climatic evidence to show that humans arrived before the Last Glacial Maximum (26.5 to 19,000 years ago) but probably dispersed during a warming period (Greenland Interstadial). Their data support the overkill hypothesis. Martin's hypothesis has gained quite a following among scientists, but there are those who disagree (Grayson and Meltzer, 2003). For example, the only two large animals seem to have been hunted were the mammoth and the mastodon. The other 34 genera were not hunted. Also, the end of the Pleistocene featured dramatic fluctuations in temperature as the Ice Age ended. The controversy continues and is unlikely to be resolved soon. Stuart (2015) strikes a middle ground by implicating both human and climate factors in the loss of the great mammals.

EARLY SPANISH SETTLERS

The first Spanish arrived in the Bay Area on November 2, 1769. Led by Don Gaspar de Portolá, they arrived overland from Mexico and claimed the area for Spain. The first Spanish settlers arrived in the Bay Area in 1776. Juan Bautista de Anza, a Franciscan priest, brought 193 colonists and about 1000 head of cattle. They built a fort called the presidio and a mission now called Mission Delores in what is now San Francisco. The population of the community averaged only 200–400, but they quickly affected the environment. Agriculture was difficult in the thin coastal soil. The cattle overgrazed the bunchgrasses (family: Poaceae), which allowed erosion to occur. Non-native crops invaded the area. The settlers cut down the trees for building material.

Early on, the Spanish decided to missionize the Native Americans by relocating them to villages near the mission so they could work for the mission. They hoped to create new colonies that would be loyal to Spain. They also tried to introduce European plants and agricultural methods, but those quickly failed. Still their efforts had a significant effect on the local environment. For example, they unintentionally brought with them European weeds that quickly took off in the New World. These included curly doc (*Rumex crispus*), sow thistle (*Sonchus oleraceus*), and redstem filaree (*Erodium cicutarium*). More recent studies have confirmed that the transformation of the California landscape occurred early and rapidly. Native grasses and weeds were quickly replaced. Native trees fared much better.

Examination of spores in mortar, material in firepits, and other evidence proximal to the missions confirms that native trees were doing well. While some smaller native plants hung on, most of the weeds, grasses, and other small plants were European. The introduced species included wheat (*Triticum aestivum*), corn (*Zea mays*), and barley (*Hordeum vulgare*). European weeds included cheeseweed (*Malva parviflora*) and redstem filaree (*Erodium circutarium*). Many of the missions were growing corn, wheat, barley, olives, and grapes.

Mission herds also grew rapidly. They included horses, cattle, swine, sheep, goats, turkeys, and chickens. Cattle and sheep eat grass down to the ground and compact the soil as they walk about, and so, their effects on the environment were significant.

RUSSIANS

In 1809, a Russian-American Company ship sailed into Bodega Bay just north of now San Francisco. They had beaver and sea otter skins, and they were looking for supplies for their Alaskan settlements. The Russians established a trading center at Fort Ross on the California coast north of San Francisco (Lightfoot et al., 1993). They also used Bodega Bay as their main harbor and had a hunting outpost on the Farallon Islands for harvesting sea otter pelts for the fur trade. The workforce comprised a few Russians and mostly Native American, Alaskan, or mixed-race individuals. At its peak, the settlement at Fort Ross probably had several hundred people.

The Northern fur seal (*Callorhinus ursinus*) existed in large numbers in the 18th Century (Jones, 2011). They spent most of their time at sea, but in spring, they come on land on islands in the North Pacific to mate and give birth. On land, they were easy to kill. Toward the end of the 18th Century, the killing of the seals increased dramatically.

By 1817, the sea otters were almost gone. From their sealing station on the Farallon Islands from 1812 to 1840, the Russians took 1200–1500 seals each year. By 1841, the fur-bearing animals were too rare to be profitable, and the Russians deserted Fort Ross. That year, the Russians sold their property to John Sutter, who would become famous in conjunction with the California Gold Rush just a few years later. The Russians' impact on Northern California was small. They seemed to have traded with the Native Americans and used them for labor, sometimes by force. However, once the seals were gone, the Russians followed soon thereafter.

EUROPEANS AND OTHERS

GOLD!

The Native Americans, Spanish, and Russians might have had little effect on California, but the next group to arrive changed everything. On January 24, 1848, gold was discovered by James W. Marshall at Sutter's Mill. The news spread like wildfire, and before long, 300,000 people arrived in Northern California. San Francisco grew from 200 people in 1846 to 36,000 just 6 years later. The roads, housing, supplies, and transportation of that great increase in population took a toll on the Bay Area. The miners lived in any type of structure, including tents and shacks.

The Gold Rush also brought destructive mining practices to the Bay Area and resulted in massive amounts of silt being washed into the Bay. Hydraulic mining was banned in 1884, but the damage was done. Although most of the silt has been washed out of the Bay in the last few decades, much of it remains in a huge sand bar just outside the Golden Gate (Fig. 6.2).

The old rivers in the Sierra Nevada contained large amounts of gold within their gravel deposits. The miners first worked the stream beds and then later moved on to the benches alongside the streams. Later, they cut into the gravel beds themselves. At first, the gold could be easily found as flakes and nuggets in streams, but that gold was quickly exhausted. The miners then graduated to small-scale placer mining by panning. By carefully swirling sand in a pan with water, the miner could use the water to wash away the lighter sand and leave the heavier gold particles. Larger-scale versions of this basic idea with sluices allowed more material to be separated efficiently. Later hard-rock mining went after gold in veins in the earth.

Soon they discovered a faster way of mining. Hydraulic mining uses high-pressure water to disrupt underground deposits. It has been used since the ancient Roman times. However, during the California Gold Rush, a more modern form of hydraulic mining introduced a nozzle that could create high-pressure water jets (Michaud, 2016). The system worked very well for extracting gold, but left behind an environmental disaster at the mine site when entire mountains were

FIGURE 6.2   Hydraulic Mining. During the California Gold Rush, hydraulic mining involves the use of high-pressure water to reduce a mountain to rubble to find the gold in it. The excess material was washed into streams and rivers and washed down into the Bay. (Credit: Carleton E. Watkins.)

washed away. Flooding and erosion increased, and streams and farms were inundated.

This form of hydraulic mining was invented by Edward Matteson. It used a hose made of canvas and later crinoline. Water pressure was achieved by using water holding ponds that were several hundred feet above the mining area. Canvas hose reinforced with iron rings delivered the water under pressure to a nozzle that varied in size from 1 to 11 inches. The larger nozzle with high-pressure water required some additional supports for handling. The technique was to aim the water stream at the base of the gravel section to "undercut" the gravel so that it would fall by gravity. A single monitor with an 8-inch nozzle could shoot 16,000 gallons of water a minute at the hillside. In a day, that one monitor could displace 4000 cubic yards of material. The amount of water was more than could be supplied by a single area. Thus, by 1859, the mining region of California contained 5726 miles of aqueducts.

This method was highly efficient. It brought far greater masses of material to be sorted in a channel that narrowed as the water and gold-rich material passed through it. This is a form of placer mining allows large amounts of material to be processed. The heavier gold particles to settle to the bottom while the water washes away the lighter impurities. It is the same principle as panning for gold, but on an industrial level.

Still, hydraulic mining is quite profitable. Relatively little hand labor is needed. The main requirements for hydraulic mining are a significant layer (about 30 feet) of gold-bearing gravel with little non-gold-bearing overburden. A ready supply of water with an adequate pressure head (a drop of 100–300 feet). For example, one mine used 60–70 million gallons of water per day. Reservoirs had to be built to store water and the run-off of melting snow, and then, the water had to be moved to where it could be used. By the mid-1880s, an estimated 11 million ounces of gold had been extracted by hydraulic mining in California. However, the cost was astronomic. Millions of tons of silt and other material had been washed into the rivers and streams that flowed into the Sacramento Valley. The rivers run more slowly over the flat valley, and the sediment was deposited in the river. The addition of massive amounts of material to the valley changed the course of rivers, increased flooding, and resulted in great costs to the re-sidents of the area.

Of course, the miners did not care. But they were soon followed by others, who were settlers, farmers, and merchants. Even in 1873, farmers began to protest that their fields were being covered by the silt that came with the floods. After a bad flood in 1875, the farmers were joined in their protests by the Central Pacific Railroad, which did not want mud on its tracks. While the farmers were relatively powerless, the railroad was a large landowner in the valley and very well-connected politically. Another landowner, Edwards Woodruff, filled suit, and after a 2-year legal battle, *Woodruff v. North Bloomfield Gravel Mining Company,* finally settled the matter. Judge Lorenzo Sawyer found that hydraulic mining was legal but dumping material into the streams was not. If the mines wanted to continue with their practices, they would have to impound the run-off.

Although hydraulic mining was allowed to continue, it quickly was abandoned. Lots of gold had been removed. Many had been made rich, and the environment would take many years to recover.

The damage caused by hydraulic mining in the Sierra Nevada during the Gold Rush was enormous. The tailings of the mining contain significant amounts of mercury. The Yuba River near Marysville, CA, was particular affected. Nakamura (2017) collected and examined samples from the Yuba Fan. He estimated that 4.24 ×103 kg of mercury are still in the floodplains of the Yuba River and, even after more than 150 years, could still affect local ecosystems.

In 1917, G.K. Gilbert completed a classic study of the sediment from hydraulic mining (James et al., 2017). Gilbert showed that sediment does not move as if on a conveyor belt from its production, storage, and transport. It moves intermittently and episodically with long periods of stasis. His work anticipated later work that showed the nonlinear movement of sediment and was the first call for what we now call environmentalism. Since that work, others have followed up to study the movement of the sediment through the Bay. For example, Schoellhamer et al. (2013) presented four models of sediment migration: a stationary natural model, a transient increasing model, a transient decreasing model, and a stationary altered model. The model covers the increase in sediment from gold mining and the decrease that began in the late 1800s. Those decreases moved into the Bay in the 1950s. Much of the remaining sediment is found behind dams and other flood control structures now. In the 1990s, more fresh water flowed into the Bay.

Sediment was not the only by-product of gold mining. Gold miners used mercury to recover gold, and more mercury was used at placer mines than any other type of gold mining (Alpers et al., 2005). Mercury contamination from these historical mines remains a potential threat to human health even today. Most of the mercury used in gold production came from deposits in the Coast Range. More than 110,000 tons of mercury was produced from 1850 to 1981, and the peak production was in the 1870s.

To recover gold from the mine sluices, hundreds of pounds of liquid mercury were added. The mercury combined with the gold to form an amalgam that precipitated. The lighter sand and gravel were washed away in the flowing water. Over time, the sluice would become coated with mercury, and other mercury was lost into the environment. The total loss of mercury during the placer mining days in California has been estimated as 5000 tons.

Inactive mines represent serious risks, and there are thought to be 39,000 inactive mine sites in California (Newton et al., 2000). About 4290 of these mines are thought to be environmental hazards, and 32,760 are thought to be physical safety hazards.

FILLING THE BAY

The Bay remained about the same for thousands of years. The arrival of the Native Americans did little to change its outlines. Even the early Europeans had little effect on the Bay. However, the discovery of gold in 1848 changed everything. Immigrants poured in the San Francisco and the Bay Area, and the wetlands surrounding the Bay were easy to fill to yield new land for development. Since the time of the Gold Rush, about 40% of the Bay has been filled, and more than 80% of the tidal wetlands have been lost.

Filling areas of the Bay continued for another 100 years. In the 1960s, some people began to become concerned about the loss of wetlands. Perhaps most notable among these was the work of three women in Berkeley. Kay Kerr, Sylvia McLaughlin, and Esther Gulick joined together to form the Save San Francisco Bay Association to prevent Berkeley from filling 4000 acres of the Bay to expand the City. However, many other projects were underway to expand cities, freeways, and airports. One project sought to take one billion cubic yards of rock and material from San Bruno Mountain and put it in the Bay to spur development in San Mateo County. Growing awareness by the public and public officials of the importance of the Bay and the wetlands began to slow these developments.

POPULATION INCREASES

The first Californians were the Native Americans, and they lived in the Bay Area for thousands of years before the first Europeans appeared. When the first Spanish explorers arrived, 200,000–500,000 Native Americans were estimated to be living in California (Mosier, 2001). The estimate for the Bay Area was 7000. The environmental impact of the Native Americans was limited. They built shell mounds, burned some areas for agriculture, and lived in relative harmony with the native animals and plants. Some anthropologists hypothesize that the arrival of humans in the Americas contributed to the extinction of many larger mammals. Tragically, 90% of the Native Americans died after the Europeans arrived.

The number of Spanish colonists in California was always surprisingly low. The average population of the Presidio of San Francisco was 200–400. It was the smallest of the three California presidios (NPS, 2015). For example, the number of Native Americans in the East Bay was far lower than the number of those who arrived from somewhere else (Arrigoni, 2020).

The discovery of gold in California changed everything. The sleepy backwater of San Francisco suddenly became the destination of thousands of people seeking their fortunes. From 1848 to the end of 1849, the population of San Francisco increased from 1000 to 25,000. The later discovery of the Comstock Lode at the end of the 1850s continued the influx of people passing through San Francisco and further fueled the growth of the city and area. San Francisco grew without any city planning, and one result is the narrow streets throughout the city. For example, the growing population required water and then sewage systems. Early on cities used private cesspools and privy chambers that had to be emptied regularly. These methods could not be scaled as San Francisco grew. As water delivery systems improved, people had more water to drink, and they did. By the 1860s, San Francisco residents were installing brick sewers. At that point, the residents were using 23 million gallons of water each day, and most of it was finding its way to the sewer system (Madrigal, 2010). The city then deployed a sewer system that channeled both wastewater and rainwater into the same pipes. Very few cities use this type of system. Rainwater tends to be cleaner than wastewater and requires less treatment. However, in the mid-19th century, sewage was not being treated anyway, and the extra water from rain just served to flush the waste and other pollutants into the Bay. Even today, in rainy weather, millions of gallons of raw and undertreated sewage are spilled into the Bay. Sewage is filled with bacteria, viruses, and other pathogens that cause various diseases in humans and are harmful to Bay plant and animal life.

The Bay Area had remained constant for a million or more years. The geological forces that created it move more slowly than humans can perceive. The first people to inhabit the Bay Area were the Native Americans, and they had little effect on the land, water, animals, and plants. The Spanish were much more involved in agriculture and domesticated animals and some rudimentary manufacturing. Yet, they also had relatively small influence on the land. That all changed forever with the discovery of gold and the influx of Europeans and others that has continued to today. In 1900, the population of the Bay Area reached 658,111, and in 2000, it was 7,150,739. In the mid and late 1800s, most of the growth was in San Francisco. The rest of the Bay Area was involved in agriculture. However, in the 1940s and 1950s, the suburbs began to build, as land was transitioned from agriculture to housing for the growing population. To support this rapid influx, infrastructure had to be built to provide housing, transportation, water, sewage, electricity, recreation, and more. That activity did not change the overall geomorphology of the Bay Area, but it did significantly change the surface. Hills were eliminated, streams were channeled, bay lands were filled, and much more. rapidly. These changes did not have the same fundamental effects of plate tectonics. However, we tend to notice them because they are so much more visible in our lives than the movement of continental plates.

REFERENCES

Allen R. (2010) Alta California missions and the pre-1849 transformation of coastal lands. *Historical Archaeology* 44: 69–80.

Alpers C.N., Hunerlach M.P., May J.T., Hothem R.L. (2005) Mercury contamination from historical gold mining in California. USGS Fact Sheet 2005-3014 Version 1.1.

Anderson M.K. (2007) Indigenous uses, management, and restoration of oaks of the far western United States. Tech. Note No. 2. NRCS National Plant Data Center, Washington, DC.

Arrigoni A. (2020) The first decades of a new era: The Native Americans of the East Bay after the Gold Rush. Museum of San Ramon Valley. https://museumsrv.org/the-first-decades-of-a-new-era-the-native-americans-of-the-east-bay-after-the-gold-rush, accessed December 7, 2020.

Bartelink E. (2009) Late Holocene Dietary Change in the San Francisco Bay Area. *Journal of California Archaeology* 1: 227–252.

Becerra-Valdivia L., Higham T. (2020) The timing and effect of the earliest human arrivals in North America. *Nature* 584: 93–97.

Grayson D.K., Meltzer D.J. (2003) A requiem for North American overkill. *Journal of Archaeological Science* 30: 585–593.

James L.A., Phillips J.D., Lecce S.A. (2017) A centennial tribute to G.K. Gilbert's Hydraulic Mining Débris in the Sierra Nevada. *Geomorphology* 294: 4–19.

Jones R.T. (2011) A 'Havock Made among Them': Animals, empire, and extinction in the Russian North Pacific, 1741–1810. *Environmental History* 16: 585–609.

Keeley J.E. (2002) Native American impacts on fire regimes of the California coastal ranges. *Journal of Biogeography* 29: 303–320.

Kiester Jr. E. (1999) Turning water to gold. *Smithsonian Magazine*. https://www.smithsonianmag.com/history/turning-water-to-gold-170493109/.

Lightfoot K.G., Wake T.A., Schiff A.M. (1993) Native responses to the Russian mercantile colony of Fort Ross, Northern California. *Journal of Field Archaeology* 20: 159–175.

Long J.W., Goode R.W., Gutteriez R.J., Lackey J.J., Anderson M.K. (2017) Managing California black oak for tribal ecocultural restoration. *Journal of Science & Technology for Forest Products and Processes* 115: 426–434.

Luby E.M., Drescher C.D., Lightfoot K.G. (2006) Shell mounds and mounded landscapes in the San Francisco Bay Area: An integrated approach. *Journal of Island & Coastal Archaeology* 1: 191–214.

Madrigal A.C. (2010) A short masterpiece on the history of sewers. *The Atlantic*. https://www.theatlantic.com/technology/archive/2010/10/a-short-masterpiece-on-the-history-of-sewers/64076/.

Martin P.S. (1967) Prehistoric Overkill, In edited by P.S. Martin, H.E. Wright Jr., Pleistocene Extinctions: The Search for a Cause, New Haven, Yale University Press, pp. 75–120.

Michaud L.D. (2016) Hydraulic mining. *911 Metallurgist*. https://www.911metallurgist.com/blog/hydraulic-mining.

Mosier P. (2001) A brief history of population growth in the Greater San Francisco Bay Region. In: *Geology and Natural History of the San Francisco Bay Area; A Field-Trip Guidebook*, edited by Philip W. Stoffer, Leslie C. Gordon (eds), *USGS Bulletin* 2188, pp. 181–186, chapter 9.

Nakamura T.K. (2017) Remains of the 19th Century: Deep Storage of Hydraulic Mining Sediment Along the Lower Yuba River, California. Master's theses. San Jose State University. 4815. DOI: https://doi.org/10.31979/etd.uvgs-92x2.

Newton G., Reynolds S., Newton-Reed S., Tuffly M., Miller E., Reeves S., Mistchenko J., Bailey J. (2000) California's abandoned mines. Department of Conservation Office of Mine Reclamation Abandoned Mine Lands Unit 1, https://www.conservation.ca.gov/dmr/abandoned_mine_lands/AML_Report/Pages/overview.aspx, accessed December 7, 2020.

NPS (2015) Spanish Period: 1776 to 1822. https://www.nps.gov/prsf/learn/historyculture/spanish-period.htm.

Schick G.W. (1994) The Ohlone and the Oak Woodlands: Cultural adaptation in the Santa Clara Valley. Santa Clara, Calif: Santa Clara University, Dept. of Anthropology and Sociology. https://scholarcommons.scu.edu/cgi/viewcontent.cgi?article=1004&context=rms.

Schoellhamer D.H., Wright S.A., Drexler J.Z. (2013) Adjustment of the San Francisco estuary and watershed to decreasing sediment supply in the 20th century. *Marine Geology* 345: 63–71.

Stuart A.J. (2015) Late Quaternary megafaunal extinctions on the continents: A short review. *Geology Journal* 50: 338–363.

7 The Bay Today

The configuration of the Bay Area was formed a few million years ago (Barnard et al., 2013). During several glacial and interglacial periods, the Bay has alternated between an estuary and a broad valley crossed by rivers that stretched more than 30 miles outside the Golden Gate. The current system that drains the Sacramento and San Joaquin Rivers through the Carquinez Strait and on into the Bay and out the Golden Gate into the Pacific Ocean was established about 600,000 years ago. The most recent flooding of the estuary occurred 10,000–11,000 years ago. By 7000 years ago, the Bay had reached its current size. Marshes began forming in the Delta about 6800 years ago. San Pablo Bay was filled by about 5000 years ago.

Millions of years of geological and other forces created the Bay. They seem to be quiet today because our timeframe is typically too short to observe them. However, there are exceptions. Earthquakes continue to occur. The most significant was the 1906 earthquake that hit San Francisco. The greatest changes to the Bay have been caused by humans themselves.

LAND

EARTHQUAKES

On April 18, 1906, a massive earthquake shook San Francisco and the Bay Area. The earthquake was on the San Andreas fault, which runs for about 750 miles from Southern California to just north of San Francisco. The horizontal displacement was 2–32 feet. At Point Reyes just north of San Francisco, the land moved north more than 15 feet. This was the largest earthquake within recorded history. However, it was not the first, and it will not be the last.

The earthquakes that occur periodically in the Bay Area are reminders that the geologic forces that made the Bay Area are still active. The mountains are still rising and the plates are still moving, but too slowly for us to appreciate. The geologic processes act on a timeline that is far longer than the existence of humans in the Bay Area. The earthquakes occur because of the relative motions of the Pacific and North American plates. The Pacific plate is moving generally northwest and the North American plate is moving generally southwest. The intersection of the two plates is called a strike-slip interaction. The plates are slipping along one another a bit at a time. Sometimes the movement is smooth. In other places, the plates become hung up on one another. Once the stress builds up sufficiently, the plates move along suddenly. The result is an earthquake.

The largest and most famous fault is the San Andreas fault that runs across most of the length of California. The slip rate on the San Andreas fault is a couple of inches per year. The general path of the fault can be seen from an airplane as one flies from San Francisco to Los Angeles. It runs through the Bolinas Lagoon near Point Reyes Station north of San Francisco through the Crystal Springs Reservoir on the peninsula just south of San Francisco and on south. The Pinnacles National Monument in Monterey County is half of an ancient volcano. The other half is 200 miles to the southeast in Los Angeles County and is called the Neenach Volcanics. The fault runs right through the volcano, and over the years, the western part has slowly moved up into the Bay Area. The San Andreas is actually a fault system with at least six more large faults across the Bay Area, including the Calaveras, Concord-Green Valley, Greenville, Hayward, Rodgers Creek, and San Gregorio Faults.

For the last couple of million years, the basic outline of the Bay has remained the same. The volcanoes are long inactive. The movement along the faults is so slow that it is not noticed by people, and the periods between massive earthquakes are too long. In terms of human experience, the geology of the Bay Area has changed little over the last 2 million years. The last big change was the water that poured into the dry Bay Area as sea levels rose after the end of the last Ice Age. That was about 10,000–12,000 years ago. Now the Farallon Islands are about 30 miles off-shore in the Pacific, and the tops of four hills remain as islands in the Bay. The largest is Alameda, and the others include Angel Island, Yerba Buena Island, and Alcatraz. An artificial island, Treasure Island, was constructed just north of Yerba Buena Island. A number of other smaller islands dot the Bay. In the East Bay, a number of extinct volcanoes can be seen. Several million years ago, the subduction of another plate, the Juan de Fuca Plate (a remnant of the Farallon Plate), under the North American plate, provided the energy for several volcanoes in the Bay Area. For the last 20 million years, those volcanoes have been extinct, and no activity has occurred in the Bay Area since then.

Still large earthquakes could and will occur on the various faults. For example, the 74 miles of the Hayward Fault passes through several significant cities in the East Bay, such as Oakland, Berkeley, San Jose, Fremont, Hayward, and others. It is parallel to the San Andreas fault. The Calaveras fault seems to be connected to the southern end of the Hayward fault, and if so, this would make the Hayward fault much more dangerous. The Hayward fault also seems to be connected to the Rodgers Creek fault to the north. Together, they could produce an earthquake of 7.2 magnitude.

As we saw in a previous chapter, the changes now have been caused by humans for the most part. Recent changes to the land mass have been caused by human activity. The shape of the Bay has changed some as much of the tidal flats area has been filled. In San Francisco, the sand dunes that once covered much of the city and the small rivers that flowed through it are all obscured by the infrastructure of the city.

The United States Geological Survey (USGS) predicts that there is a 72% chance that an earthquake of 6.7 magnitude or more will occur somewhere in the

Bay Area before 2043 (Aagaard et al., 2014). Extensive development has also taken place all around the Bay so that the land now supports the 5th largest metropolitan area in the United States with a population of 7.75 million people. An earthquake of that magnitude will be devastating.

OTHER LAND MOVEMENT

Earthquakes are not the only ways that earth, soil, and mud move. Mud and silt flow downhill, and humans have influenced that for thousands of years. For example, clearing land for crops encourages erosion that fills rivers and streams. In more recent times, deforestation and urban development have resulted in more erosion, while dams have stopped the natural movement of the material (Voosen, 2020). Landslides cost lives and damage property.

New space-based imaging technologies, such as synthetic aperture radar interferometry (InSAR), have provided significant improvements in measuring surface deformations that can be informative to examining faults, ground movements up and down, landslides, and more. Uplift results from the movement of the plates along the faults. Downward movements can result from depletion of aquifers and settling of unconsolidated sediments. Landslides also cause land to move.

Ferretti et al. (2004) used InSAR to look at the movement of land in the Bay Area. They noted an uplift of 0.4 ± 1 mm/year in the East Bay between the Hayward and Calaveras faults. Other areas, such as San Leandro and the Santa Clara Valley, demonstrated much more rapid uplift, probably due to increases to confined aquifers. In still other areas, erosion and landslides have reduced the hills by as much as 27–38 mm/year in the Berkeley Hills. The greatest amount of subsidence was seen in areas where building occurred over Bay mud and fill. Parts of Treasure Island and Alameda and southeast San Francisco dropped by about 0.15 meter in 1992–2000.

Slow-moving landslides are a problem throughout much of the Bay Area and particularly in the East Bay hills. They rarely cost lives, but they can be extremely destructive to structures and infrastructure. The slides depend on the soil, climate, and earthquake activity of the area, but the mechanisms that initiate and maintain them are not well understood. They often occur in soils rich in clay and rock that are mechanically weak and have high levels of seasonal precipitation. Cohen-Waeber et al. (2013) used a network of global positioning stations (GPS) and InSAR to measure surface displacements with extreme accuracy. The area that they examined is part of the California Coast Range, which is a mix of Jurassic to Tertiary sedimentary, volcanic, and metamorphic rocks. It is overlain with Quaternary colluvial and alluvial deposits. Importantly, this soft rock has been subjected to considerable fracturing and weathering and is susceptible to sliding. They found that the land movements and velocities were quite similar. They found only limited influence of earthquakes, even though some small to moderate earthquakes occurred during the measurements. Rainfall seemed to be a key element in the initiation of landslides.

The California Coast Ranges are an ideal place to study slow landslides. Lacroix et al. (2020) reviewed the forces that control slow landslides in this environment. Those factors include the overall geology, climate, and tectonics and also precipitation and groundwater, earthquakes, river erosion, anthropogenic activities, and external material supply. Rivers can block slow-moving landslides, and landslides can block rivers. Finnegan et al. (2019) examined the several effects on the blockage of rivers by landslides and found that wider rivers are less affected by landslides. More narrow streams can be completely blocked by the landslide. Also rivers vary in their ability to mobilize the material in the landslide.

In another study, Schmidt and Bűrgmann (2003) used InSAR data to show uplift in the Santa Clara valley from 1992 to 2000. The uplift was 41 ±18 mm. Furthermore, they observed seasonal uplift and subsidence west of the Silver Creek fault. The fault may act to block the movement of underground water. Water in the Santa Clara valley aquifer moves from one area to another. The flow is channeled by the nature of the sediments in specific areas. Flow is rapid through sand and gravel, but slower through clay. Drawing down the aquifer disrupts the flows in the aquifer. Since landslides are often associated with extreme weather events, Cordeira et al. (2019) compared landslides from 1871 to 2012 with records of Pacific winter storms and atmospheric river events. They found that 76% of the landslides occurred during storms and 82% occurred during an atmospheric river.

LAND USE

While the basic structure of the land in the Bay Area has not changed significantly over the last 2 million years, its use has. The bulk of the land has changed from wild lands to farm lands and more recently to urban infrastructure. The Bay Area has 2.3 million acres of farmland in its 3.6 million acres of greenbelt. However, over the last 30 years, 217,000 acres of agricultural land was repurposed to urban growth, and 200,000 acres are at risk, according to the Greenbelt Alliance.

WATER

REGIONS OF THE BAY

The Bay is a complicated ecosystem with at least three somewhat different regions. The North Bay receives more than 95% of the Bay's freshwater input. Gravity causes the freshwater to flow toward the sea and salty water to flow toward the Delta. The mixing from this action creates a gradient of salinity that increases as one gets closer to the Main Bay, which is the most salty region. The South Bay receives little freshwater and much of that is treated wastewater. There is so little inflow that there is little mixing due to gravity. In fact, for much of the year, the South Bay is somewhat isolated from the rest of the Bay and is more like a tidally oscillating lagoon.

Thus, the Bay can be examined more carefully by considering the several regions that make it up. They differ in salinity, exposure to wind and waves, plants and animals, pollution, and more. We will briefly introduce the Bay regions and later in the chapter describe aspects of them in greater detail.

The Golden Gate and the Main Bay

At the mouth of San Francisco Bay lies an ebb-tidal delta that has evolved over the last 130 years. Erosion and deposition compete to diminish or add to the delta, and human activities influence both. As water flows out of the Bay, it carries with it sand and other debris that is deposited into the delta. Those amounts vary with the amount of water that leaves the Bay, which is called the tidal prism. The tidal prism is calculated as the volume of water in the Bay between mean high tide and mean low tide. In other words, it is the amount of water that leaves the Bay at ebb tide. The more water that leaves the Bay, the greater the tidal prism, and the greater the amount of sand that is deposited in the delta.

In 2015, new maps showed the underwater landscape just outside the Golden Gate (Johnson and Gordon, 2015). Strong tidal currents have given this area several interesting features, including sand waves, a deep scoured area directly at the Golden Gate with a sand bar beyond it, and the "Potato Patch" shoal that can be a hazard to navigation by small craft. Ocean Beach and the coast for about 30 miles south of the area are a hotspot for beach erosion that will be more of a problem with ocean rise due to climate change.

Dallas and Barnard (2011) analyzed four historic bathymetric surveys of the San Francisco Bar conducted between 1873 and 2005. They found that the sand bar had lost an average of 80 cm (equivalent to about 100 million m^3 of sand) and contracted so that the main mass is now about 1 km closer to shore. These findings are in agreement with the earlier work that showed a 9% loss of the tidal prism. Human activities have affected the total levels of sediment, including filling large parts of the Bay and diking the rivers that feed it. In addition, large amounts of material from the Bay has been removed by dredging or mining for sand and aggregates for construction. Perhaps 200 million m^3 of material were removed from the Bay system. Loss of this mass of material has significantly accelerated tidal delta and open beach erosion in and near the Bay.

South Bay

The most southern section of San Francisco Bay is mostly quite shallow (3–6 m). The deepest section is about 26 m at Hunter's Point. It receives freshwater input from the Guadalupe River. A main channel runs down the approximate center of the South Bay (Foxgrover et al., 2004). It is about 1 km wide, and its depth is 11 m. The channel is surrounded by much shallower shoals. The area south of the Dumbarton Bridge has remained quite stable for the last 125 years at least.

Foxgrover et al. (2004) examined five hydrographic surveys that were conducted from 1858 to 1983 by the National Ocean Service and its predecessor the US Coast and Geodetic Survey. By reconstructing the floor of the Bay at each time point, they could determine how the South Bay has changed over those 125 years. They combined modern (e.g., digitized soundings) and traditional (hand-drawn maps) methods and corrected for land subsidence that occurred between 1934 and 1967. They found that the South Bay has had alternating periods of deposition and erosion. However, the Bay has lost about $90 \times 10^6 \, \text{m}^3$ of sediment. Strikingly, 80% of the tidal marsh was converted to salt ponds, agricultural uses, and urban areas from 1858 to 1983, and overall, 40% of the intertidal mud flat area was lost. The loss of tidal flats was particularly severe on the eastern shore north of the Dumbarton Bridge.

The dominant feature of the region is the water in the San Francisco Bay. The Bay is an estuary that comprises the Bay, San Pablo Bay, the Carquinez Strait, Suisun Bay, the Sacramento-San Joaquin River Delta, and the Guadalupe River and covers approximately 1600 square miles. Nearly 40% of the precipitation that falls on California eventually finds its way to the Bay.

The Bay is actually surprisingly shallow. The average depth is about 14 feet, and it is constantly changing due to tidal currents, dredging mining for aggregate, and landfill. The deepest part is at the Golden Gate, where strong tidal flows have scoured the bottom to a depth of 350 feet. Between Alcatraz and Angel Island, the Bay is 100 feet deep. At the southern end, it is only a couple of feet deep.

The Bay is subject to the tides from the Pacific Ocean. Tides change every 6 hours. As it comes in, 640 billion gallons of water flow under the Golden Gate Bridge into the Bay, and the surface of the Bay increases by 50 square miles, and the mudflats disappear for a while.

North Bay

The North Bay includes San Pablo Bay, Suisun Bay, and the Sacramento River Delta. Most of the freshwater that enters the Bay does so from the Delta. In the past, 20–24 million acre-feet of water came into the Bay that way. Over the years, various projects have redirected 30–40% of the water elsewhere. The amount of water varies from high levels in years of El Niño to much less when the state is in drought.

Suisun Bay is the northern-most part of the San Francisco Bay system. It lies at the entrance of the Sacramento and San Joaquin Rivers into the Bay and the entrance to the Sacramento Delta. The Army Corps of Engineers maintains a shipping channel that is 300 feet wide and 35 feet deep (Mean Lower Low Water) from the Carquinez Strait at Martinez to Pittsburg and a channel 250 feet wide with a depth of 20 feet further up to Port Chicago.

The topology of Suisun Bay was explored by five hydrographic studies between 1867 and 1990. A surface map was prepared for each survey, and the maps were compared (Cappiella et al., 1999). Cappiella et al. (1999) determined the volumes of erosion and deposition, sedimentation rates, and shoreline changes. Hydraulic gold mining resulted in very large amounts of sediment in the Bay (115 million m^3) between 1867 and 1887. During that period, two-thirds of the Bay was receiving material and one-third was losing material due to erosion. However, after that and until 1990, the area overall lost material. Suisun Bay changed from an area of deposition to one of erosion. Deposits were particular large during the period when hydraulic mining was underway in the gold fields in the late 19th century. Hydraulic mining stopped during the earlier period, and that meant that less material was being transported down the rivers to the Bay. In addition, measures to control floods and distribute water lowered the sediment in the rivers. Finally, area of the tidal flats was greatly reduced.

The water in Suisun Bay eventually flows into San Pablo Bay. The Bay is about 90 square miles and is relatively shallow, but it has a deeper navigable channel that runs through the middle of the Bay. Sediment from Suisun Bay works its way into and eventually through San Pablo Bay and on to the main Bay.

Higgins et al. (2007) mapped the bottom of San Pablo Bay by examining earlier studies. They found a similar story to that of Suisun Bay. Mining debris in the late 19th century pouring into the Bay. Several cycles of deposition and erosion followed. That debris and other material were deposited onto the mud-flats and shoals and later redistributed and eventually washed on into the main Bay. They also examined the radioisotope profiles of the sediment throughout the Bay. In general, those results are in agreement with the earlier studies. More specifically, they studied the levels of ^{137}Cs and ^{210}Pb in 12 core samples of sediment.

SEDIMENT

The Bay is a complex system. Human activities, including mining, dredging, reservoirs, freshwater diversion, urban runoff, shipping, exotic species, and attempts to restore wetlands, have added to that complexity. The outlet for all of the water and its pollutants is the Golden Gate, and it is subject to tidal movements. It is important to remember that the various parts of the Bay (e.g., the Golden Gate, main, north, and south bays, and the Delta) are all interconnected into a large system (Barnard et al., 2013). Understanding how the many processes work in the Bay requires looking at the system as a whole.

Sediment particles can be either organic or inorganic, and they vary greatly in size, shape, and composition (Kjelland et al., 2015). They are important because they contain nutrients and sometimes contaminants. Sediment results from natural causes (e.g., erosion, freeze-thaw cycles) or human activities (e.g., mining, agriculture, construction). They are suspended and carried by the actions of water flows. At some point, they settle to the bottom until disturbed by greater water flow, action of an animal or humans, or other causes. Sediments are often classified according to how they are transported. Larger particles, called the bed load, move along the bottom of the waterway. Smaller particles (e.g., clays and silts), called the suspended load, are suspended in the water and move with the water flow.

Suspended particles affect the clarity of the water and absorb heat energy and raise the temperature of the water. They also block sunlight from penetrating deeper into the water, and in that way, they deter photosynthesis and the production of oxygen by green plants. Thus, sediment can have significant effects on plants and animals.

Much of the sediment arrives via the rivers that feed the Bay. Those rivers flow into a large delta of about 3000 km^2 involving many smaller rivers and streams. These factors make it difficult to estimate the flow and amount of sediment entering the Bay.

The organization of an estuary takes decades or centuries or longer to establish. Even then, the sediment does not stay in the same place even after it is deposited. It is transported by tides, wind, animals, and human activities. The amount of sediment is determined by a balance of the newly arriving sediment and the loss of sediment out of the estuary. Various studies have attempted to define this balance in San Francisco Bay over different periods. Moftakhari et al. (2015) estimated that 55% of the 1500 ±400 million tons of sediment delivered to the Bay from 1849 to 2011 was due to human activities. Furthermore, they believe that the amount of sediment being delivered to the Bay now is 50% less than at its peak. Jaffe et al. (2007) looked at records of the sediment in San Pablo Bay from 1856 to 1983. In the earlier days, the sediment patterns were more complex. By 1983, all of the earlier channels and deltas were gone. Only the main channel remained. The rest was a gentle grade from the shore to the main channel. The amount of material entering the Bay has also changed. During the years of hydraulic gold mining, 14 million cubic meters of material arrived in the Bay. The intertidal mudflats grew by 60%. After 1951, dams and other flood control efforts greatly reduced that amount of sediment reaching the Bay, and the material that was already there began to be washed out. The mudflats reverted to the levels they were before the mining began. McKee et al. (2006) looked at the suspended sediment load between 1995 and 2003 arriving at Mallard Island. Mallard Island is at the mouth of the Delta where the Sacramento and San Joaquin Rivers flow into the Bay. They estimated that the peak annual amount of sediment was about 1.2 million metric tons. About 20% of the total was dispersed toward the land. The rest was transported on to the Bay.

The sediment in the Bay already and the newly arrived sediment every year are a mixed blessing. On the one hand, the sediment provides habitat for many plants and animals. On the other, it is a problem for shipping. The Port of Oakland is one of the busiest on the West Coast. It handles over 2 million containers per year. Moreover, most of the Bay is shallow, and large amounts of slit are deposited each year. Thus, the shipping channel must be regularly dredged. From 1995 to 2002, 3.1 million cubic meters of material was dredged from the Bay. For some time, much of the dredged material was deposited further out at sea. Some of it was used to reconstitute wetlands that had been depleted by the normal washing of material out of the Golden Gate.

Dredging stirs up the sediment already in the Bay and has complex and not completely understood effects on the animals and plants. Organisms close to the dredging operation or in a sensitive stage of life are affected most. Some can move away from the plume of unrooted sediment, but others cannot. The long-term effects of exposure to the suspended material.

The hydraulic gold mining of the late 19th century washed enormous amounts of sediment down the rivers that feed the Bay. Even the southern end of the Bay was affected. Other human activities have continued to change the South Bay. For example, while erosion is a major natural factor in loss of sediment, removal of material by human activity has also been significant. Dredging began in the Bay in the 1880s to clear the harbors for shipping. In addition, a great deal of material was removed for filling areas along the shore for construction and other reasons.

The amounts of sediments reaching the Bay have been greatly reduced over the last decades. The Gold Rush ended over 100 years ago. Those massive amounts of sediments have been washed out of the Bay, and more controlled rivers allow less material to be eroded into the Bay. Perhaps half of the current sediment load comes from the bedrock that underlies the local watersheds (Elder, 2013).

WATER CHARACTERISTICS

Dissolved Oxygen

The amount of oxygen in water is critical for the animals that live in the water, and it is related to the nutrients and chemicals in the water. For example, bodies of water that have large amounts of runoff of wastewater or agricultural fertilizers tend to have large blooms of algae and other plants that consume much of the available oxygen in the water. The lack of oxygen affects the survival of fish and other animals. The Lower South Bay region experiences this problem, particularly in the sloughs that carry discharged water (MacVean et al., 2018). The amount of oxygen varied with the tides. Periods of low oxygen levels were the shortest at the Dumbarton Bridge and longest in the Guadalupe Slough. The levels of oxygen were also examined in terms of other characteristics (e.g., temperature, clarity, and salinity) and, most importantly, of the fish that could be supported in those areas. Over 50 species of fish were surveyed in the study. Interestingly, the abundance of fish was greatest in areas of low oxygen. They speculate that this might represent a concentration of fish that do well in those conditions. Temperature and salinity were the greatest variables for most species.

Salinity and Temperature

The San Francisco Bay estuary is an important ecosystem that provides a home or a resting place for migrating species. The salinity and temperature are critical factors that affect the vegetation and animal species in the Bay and are affected by multiple factors, including the activities of humans.

The salinity in the Bay determines the plant and animal life in the estuary. That salinity is a balance of the flow of freshwater into the Bay from the rivers and the tidal movements and baroclinic flows that bring saltwater in through the Golden Gate and varies as those influences vary. At steady state, they are in balance, but flows or tides change them constantly, and the levels of salinity change in estuaries. Freshwater influx from rivers and streams lowers the salinity with freshwater. The actions of tides and wind on seawater and dispersion from baroclinic exchange flows are the main variables for transporting salt upstream in the estuary. Precipitation, evaporation, and groundwater are other more minor factors determining salinity. The salinity of the Bay also changes seasonally. Winters have traditionally been wet in the Bay Area, and so, the inflow of freshwater increases during those periods of heavy precipitation. In the summer, the amount of freshwater is much less. Also a number of the small streams that feed the Bay are dry in the summer months and only fill with water during the rainy winters. Salinity varies according to the season. For example, according to the USGS, the salinity of the Southern San Francisco Bay at a point just south of the San Mateo Bridge changes from about 93% in the dry summers to about 50% in the wet winters.

The measure of salinity is the practical salinity unit (psu). It is based on the conductivity of seawater, and 1 psu equals 1 g of total salt per 1 kg of seawater. Average seawater is 35.5 ps. However, seawater salinity varies: freshwater is near 0 psu; water near the mouths of rivers is about 15 psu. The Dead Sea is 40 psu. The ocean water in Northern California is about 33 psu. The Bay averages about 11 psu. Thus, the Bay is a mixture of about two parts seawater and one part freshwater. Monismith et al. (2002) examined the salinity of the Northern San Francisco Bay. They used the EPA measure of X, which they defined as the distance (km) from the Golden Gate along the main shipping channel to a point where the salinity at the bottom was 2 psu. That point moves throughout the year, but it is an excellent indicator of salinity.

Obviously, the major source of saltwater is the Pacific Ocean. Freshwater comes from the major rivers that flow into various parts of the Bay. The two largest are the Sacramento and San Joaquin Rivers, but others are the Napa River and Petaluma River. The Guadalupe River drains into the Southern Bay. The ultimate source of the freshwater is rain and snow that falls in the Sierra Nevada Mountains. Finally, numerous other smaller streams empty directly into the Bay at different points.

Ingram et al. (1996) studied the salinity and water flow in the San Francisco Bay. To do this, they measured levels of oxygen and carbon isotopes in fossil bivalves in the sediment. From 1900 to 1670 years before today, they found that the amount of water flowing into the Bay in that period was significantly greater than before diversion began. However, between 1670 and 750 year ago, there was a great reduction in the flow. Since then, the flow has fluctuated between wet and dry periods.

The National Oceanic and Atmospheric Administration (NOAA) maintains a series of stations throughout the Bay that track salinity and other factors in real time called San Francisco Bay Operational Forecast System.

Salinity in the Bay is a major factor in the health of the plants and animals. The Delta is a particularly important area. It is referred to as a nursery for many important species. Seasonal changes in salinity can significantly affect these species, and the human impacts on salinity must also be considered by fishery managers. Finally, salinity can be an important marker for tracking toxic materials in the Bay.

The temperature of the water of the Bay is an important factor. Ocean waters near are warmest from mid-August to late September and average 60°F. They are coolest in January and February, when the water is typically 53°F. Vroom et al. (2017) developed a model of temperature for an estuary and tested it in the San Francisco Bay. They found that, in the North San Francisco Bay, temperature is related to salinity stratification. The temperature and amount of water from the rivers feeding the Bay and influence of the atmosphere affect water temperatures differently throughout the Delta.

The water in the Pacific off the Bay tends to be cold. The Coriolis effect, related to the Earth's rotation, causes the northwesterly winds to push the surface waters along the coast away. Colder water from deeper levels replaces the surface water. The cold waters make the waters near the Bay create a nutrient-rich environment. They also cool the air above the water, and when that cooler air hits the warm air over the land, the result is San Francisco's famous fog. The upwellings occur mostly in the spring and summer.

Cloern et al. (2007) noted an increase in the phytoplankton mass beginning in 1999. They were surprised by the increase because the typical indicators of an increase were missing: nitrogen and phosphorous levels were lower than normal, and algal blooms usually follow an increase in these compounds. The nitrogen and phosphorous result from agricultural runoff. Other changes involved a reduction in the number of bivalves and an increase in the number of bivalve predators. These changes resulted from a cold phase in the Eastern Pacific Ocean that affected the Bay waters. The colder waters brought an influx of predators into the Bay that lowered the population of bivalves and, thus, allowed the phytoplankton levels to increase.

WATER QUALITY

San Francisco Bay is extraordinarily beautiful. The hills surrounding the Bay offer amazing vistas of sailing boats and ships, cities, and graceful bridge. Fog is common, especially in the summer, and adds to the romance. Beautiful as it is, the Bay is an enormous and complex system of tides, temperature, salinity, silt, pollutants, animals and plants. The health of the Bay is not measured simply in the quantity of water. It is also a matter of the quality of that water. Bay water has changed greatly over the many years. For most of its existence, the quality was a function of natural processes. Increased rainfall made the Bay less salty. Silt washed down the rivers and then out the Gate. In the last 100 years or so, humans have been responsible for even more radical changes. As Cloern and Jassby (2012) note, "The Bay moderates regional climate, assimilates wastewater from 50 municipal sewage treatments plants, is a center of commercial shipping, serves as both nursery and migration route for ocean-harvested fish and crabs, and includes the largest tidal wetland restoration project in the western United States."

The Bay is an estuary and so is influenced both by the upwellings and movements of the Pacific Ocean at its mouth and the addition of freshwater, primarily from the Sacramento and San Joaquin Rivers. Winters are wet and summers are dry, and the runoff varies greatly from a low of 7.6 km^3 in 1977 to a high of 65 km^3 in 1983 (Cloern and Jassby, 2012). The average annual runoff is somewhere between 20 and 35 km^3.

Freshwater is critical to an estuary (Kimmerer, 2002), and the amounts of freshwater entering affect the estuary in many ways. Clearly, the salinity of the estuary rises and falls with the amount of freshwater entering. Seawater is quite salty, but that salinity can be moderated by adding freshwater. The freshwater helps to dilute pollutants in the estuary. It also adds to the complexity of the estuary by enhancing the stratification. Finally, it tends to wash materials out of the estuary and shorten the time they spend in the Bay. All of these are healthy and important effects on the Bay.

Pollution

Most pollution in the Bay comes from the local watershed that drains into the Bay. The USGS has been measuring the water quality in the SF Bay since 1969. Their measurements include salinity, temperature, light extinction coefficient, and concentrations of chlorophyll-a, dissolved oxygen, suspended particulate matter, nitrate, nitrite, ammonium, silicate, and phosphate. The data collected by USGS have been used in many studies, and the results were summarized by Schraga and Cloern (2017).

Sediment is an important factor in the health of the Bay. For example, many contaminants associate with sediment particles (Schoelhamer et al., 2007). Thus, measurements of sediments are a useful surrogate for other contaminants. The Regional Monitoring Program for Trace Substances uses this strategy to monitor a number of trace metals and organic contaminants. The suspended sediment concentration varies with over 2 weeks, a month, and 6 months. Sediment flows into the Bay and is resuspended by the winds and waves, and the fine sediment is eventually washed away. Contaminants in the sediments are concentrated in shallow water.

Plants and animals often concentrate contaminants in their tissues. This action can make fish and other seafood dangerous to eat, but it also provides another method for following the levels of various contaminants. Greenfield et al. (2005) examined contaminants [i.e., mercury, selenium, polychlorinated biphenyls (PCBs), dichlorodiphenyltrichloroethanes (DDTs), chlordanes, and dieldrin] in sport fish in the Bay. The samples were all above recommended levels for all contaminants except chlordanes. The levels correlated with the species of fish, year, and amount of lipid in the fish. Longer-lived and larger fish (e.g., leopard shark, striped bass, and white sturgeon) had higher levels of mercury. Fattier fish (e.g., white croaker and shiner surfperch) had higher levels of PCBs, DDT, and chlordane. However, they saw no long-term trends.

Sediment is the vehicle for 97% of the loading of transition metals into the oceans (Gibbs, 1977), and mercury is one of the most serious of these contaminants. The USGS found that it was widespread across the western United States (Eagles-Smith et al., 2016). They found it in the air, soil, sediment, plants, fish, and wildlife. The most serious risk for humans is in eating contaminated fish. The movement of mercury in from the soil to plants and animals is complicated and involves interactions between the soil, atmosphere, plants, and water. Mercury is found either as inorganic mercury mainly in the soils or as methylmercury often in fish, birds, and other living things. Inorganic mercury is converted into methylmercury by the action of bacteria in water. Interestingly, the authors found that most inorganic mercury tends to be localized. They suggest that the key to controlling mercury in the food chain is to prevent the inorganic mercury from being converted into methyl mercury.

Donovan et al. (2013) reported the levels of mercury around the Bay that they found in sediment cores and found that they are what one would have predicted, knowing the history of the area. More specifically, they compared the levels of total and different mercury isotopes (i.e., ^{199}Hg and ^{202}Hg) from pre-mining uncontaminated sediment, sediment from around 1960, and surface sediment. The pre-mining era sediments have less than 60 ng/g of mercury. The surface levels are similar to those on the Yuba River and are probably from the Hg used in the gold mining in the Sierra Nevada. The deeper sediment layers in the wetlands in the south and central SF Bay have high total Hg levels (>3000 ng/g). These probably result from human activity.

More specific to the San Francisco Estuary, Greenfield and Jahn (2010) examined the amounts of methylmercury in small forage fish from 2005 to 2007. Mercury concentrations were higher in fish species that inhabited the mudflats and wetlands (e.g., *Clevelandia ios*, *Menidia audens*, and *Ilypnus gilberti*) than in those that lived offshore (e.g., *Atherinops affinis* and *Lepidogobius lepidus*). Interestingly, *Atherinops affinis* and *Menidia audens* had similar diets. Finally, the levels of mercury were higher near the inlet of the Guadalupe River, which traditionally has had significant mercury contamination from upstream mining activities.

Unfortunately, the mercury contamination in the Bay is likely to persist for some time into the future. Much of the mercury contamination from the gold mining period is still in the hills above the rivers. Singer et al. (2013) studied the movement of mercury from the Yuba Fan to the Bay. Large floods periodically wash that mercury out to begin its migration to the Bay. From their analysis of the data and their model, they estimate that it will take 10,000 years to remove all of the mercury. Thus, the legacy of large-scale mining persists far after active mining has ended. This research has far-reaching implications for future mining or other human activities that disturb large areas of the natural environment.

Polycyclic aromatic hydrocarbons (PAHs) are organic molecules that contain multiple aromatic rings, such as naphthalene and anthracene. They are found in coal and tar and can be produced by the burning of organic matter, such as in a forest fire. PAHs are ubiquitous in nature. They are formed under high temperatures and, thus, are typically mixtures of different molecules rather than a pure sample. They also result from human activities, such as smoking. PAHs have been linked with several types of cancer in humans and also with cardio-vascular disease and poor fetal development. Their association with cancer is long-standing. They are thought to be the cause of the increased risk of scrotal cancer in chimney sweeps in 18th-century England. For these reasons, they are regulated by the US Environmental Protection Agency.

PAHs are very large nonpolar molecules. Two- (e.g., naphthalene) and three-ring (e.g., anthracene) PAHs are soluble in water, and two- to four-ringed PAHs are volatile and evaporate into the atmosphere. Larger molecules are not volatile at normal temperatures and not soluble in water. However, they do attach to fine particles of organic matter in water. That is a two-edged sword. Those large PAHs are not that biologically available, but they remain in the environment for a long time.

PAHs get into rivers and bay when rain runs off roads, through sewage, and particles in the air falling to the surface. Some industries, such as creosote manufacturing, and coal-burning power plants can have very high levels that can be transported to waterways. Algae and invertebrates (single-celled organisms, mollusks, and some worms) accumulate PAHs in their tissues. Mussels can be particularly affected. Interestingly, vertebrates seem to more efficiently excrete PAHs than the lower forms. PAHs are slowly degraded by sunlight and other actions, especially near the water's surface. However, some of the metabolites can be more toxic than the original PAH.

Amounts of PAHs in waterways vary according to the human activities in the area. Cities and industrial activities clearly increase their concentrations. Tides near the estuary mouth dilute their concentrations.

PAHs are common in the Bay. The numbers are startling. Estimates are that 10,700 kg/year enter the Bay (Oros et al., 2007). The major sources are pretty much the same here as elsewhere and include storm water runoff, effluent from wastewater treatment, particles from the air, and dredging. Since 1993, the Regional Monitoring Program for Water Quality has monitored their levels in the water, sediment, and mussels. From 1993 to 2001, there was little change in their levels by season, but a number of monitoring stations reported levels higher than recommended. Unfortunately, they conclude that the levels in the Bay are not likely to be reduced in the near term.

Sewage

One of the most serious was raw sewage (for an excellent review, see Cloern and Jassby, 2012). Complaints of foul smells and sights spurred action in the 1990s. The problem was not just esthetic. Sewage is not just smelly and unsightly. It contains pathogens, such as bacteria, viruses, and more. Fish and shellfish populations were affected by the resulting anerobic conditions. These were especially notable near sewage outfalls in the East and South Bay. Over time, treatments improved, and the outfalls were moved farther out into the Bay so that the contents could be better diluted. Oxygen levels are less affected now. The South Bay has been particularly affected by sewage discharge. It has the highest levels of dissolved inorganic nitrogen of any estuary in the United States. Levels of dissolved inorganic phosphorus are also high.

However, sewage spills and overflows still occur, especially in times of heavy rain. In 2010–2011, nearly 250 million gallons of sewage and contaminated rain were released into the Bay. The additional rain simply overwhelms the ability of the plants to treat the waste. The plants are out of date and inadequate to the excessive demand. Rainwater is meant to be captured by the storm sewer system, not the treatment plants. However, the water seeps into the system through broken pipes and connections. The pipes can become blocked so that they back up and spill sewage in streets or allow it to be flushed into the Bay. The City and County of San Francisco is an exception. Their sewer system is designed to collect both sewage and rain runoff. In support of their system, they point out that they are able to treat both and discharge cleaner water into the Bay than most other municipalities.

There are some hopeful signs. In 2014, the East Bay cities of Oakland, Berkeley, Alameda, Albany, Emeryville, and Piedmont and the East Bay Municipal Utility District agreed with the US EPA to spend US$1.5 billion to replace 1500 miles of aging sewer pipes to improve the discharge of treated water into the Bay after heavy rains. The Bay Area is a world-class port for shipping. As of 2012, ships of more than 300 tons are not allowed to discharge sewage or gray water into California waters. All ships entering California waters must be inspected by the US Coast Guard. The Environmental Protection Agency estimates that this rule prevents the discharge of 23–25 million gallons of treated sewage into California waters each year.

Microplastics

Plastics have become a major source of pollution in marine environments (Auta et al., 2017). The same is true of estuaries. In recent years, microplastic fragments smaller than 5 mm have been recognized as a significant problem (Masura et al., 2015). Sutton et al. (2016) examined microplastics pollution in the San Francisco Bay. They found an average of 700,000 particles/km^2. The Bay surface water has more microplastics than any other body of water in North America. Even the treated wastewater is contaminated.

Microplastics are insidious because they come from so many everyday products and processes. They come from microbeads in personal care products, the manufacture of plastic products, fibers from fabrics and fishing line, photodegradation of plastic items, cigarette filters, and more. For example, the mechanical and chemical stresses of washing synthetic textiles releases a large amount of microplastics. De Falso et al. (2019) attempted to quantify the yields. They processed clothes in household washing machines and examined the effluent. They found that 124–308 mg/kg of fabric were released. This corresponded to 640,000–1,500,000 particles. The size of the particles was such that they would pass through the filters in wastewater treatment plants. The problem with them is that they can enter the food chain in many ways. They can physically block the digestive tract and cause other problems. Worse still, the microplastics accumulate organic pollutants, including PAHs.

Microplastics are a particular threat to marine ecosystems. They originate from activities on land, but they are washed into the water by runoff and even from water treatment plants. The concentrations released by water plants are very small, but the amount of water processed in any city is staggering. Daily discharges can be millions of particles (Mason et al., 2016). Fibers and fragments are the most common types of particles. Worse still, many personal care products contain microbeads, and billions of these particles are released into river, streams, and bays every day in the United States.

Pollution at Former Military Bases

In the 1990s, several military bases in the Bay Area were closed, including Mare Island, Hunters Point, Oakland Army Base, and Concord Naval Weapons Station. Before ownership of the properties could be transferred to cities or other entities, each had to be cleaned up. Unfortunately, the bases had been in existence since long before environmental regulations were implemented. Many parts of the bases had been contaminated by the dumping of a broad range of chemical and radioactive waste (Semler and Blanchard, 1989). Treasure Island contained large amounts of lead, dioxins, petroleum products, and more than 1000 radioactive items (Respaut and Levinson, 2019). Hunters Point Naval Shipyard was declared an EPA Superfund site (EPA, 2020).

CHANNELING AND FLOOD CONTROL

Filling of the Bay

Until the mid-19th century, San Francisco Bay was surrounded by extensive wetlands. In the center of the Bay, there was a deep channel that followed the path of what had been a river before the inundation that occurred about 20,000 years ago. During the era of hydraulic gold mining, massive amounts of rock and debris were dumped into the Sacramento and San Joaquin Rivers and transported down the rivers to the Bay. The results were that much of the shallow areas of the Bay were filled in and new wetlands were created.

San Francisco sits at the tip of a peninsula and is surrounded by water on three sides. The amount of land is limited by this geographic fact. Yet the demand for additional land increased dramatically. Filling the wetlands seemed like a great solution to this problem, and over the years, the shoreline of San Francisco was greatly expanded by pumping or dumping dredged or Bay material on the wetlands. These areas were created with little care, and now, they are areas of great concern for potential liquefaction during earthquakes.

These include some very expensive areas in San Francisco, such as Bayfront, the Marina district, the financial district, and SoMa (south of Market). Mission Bay, now the site of the new basic sciences campus of the University of California, San Francisco, was a small bay in the middle of a large marsh. It and other areas were used as a dump for debris left over from the 1906 earthquake and fire.

Since then, humans have made other changes to the Bay. Around the turn of the 20th century, the US Army Corps of Engineers began dredging the rivers and the Bay to make them deep enough for shipping. The Bay is actually quite shallow. Most of it is less than 15 feet deep, and that part south of the San Mateo Bridge is only a few feet deep. Many large ocean-going ships need a channel of 50–55 feet. Some of the dredged material was used to build Treasure Island in the 1930s. In addition, many of the wetlands were filled to expand the land available for building. In recent times, the dredged material has been used to re-establish wetlands around the Bay.

SEA LEVEL RISE

About 10,000 years ago at the end of the last Ice Age, the melting ice caused sea levels around the world to rise. In the Bay Area, the seawater eventually rose high enough to flow over the Golden Gate and flood the plain that we now recognize as the Bay. For the last 10,000 years, sea levels have risen only slowly. During the last century, the sea level has risen an average of 1.96 mm/year for a total increase of 7.7 inches (Curry, 2018). However, the world is now in the midst of another global warming scenario, and sea levels are beginning to rise again.

Also over the last 10,000 years and particularly in the last 200 years, humans have filled parts of the Bay to obtain new land for development. Those areas are still unstable and tend to be lower than other areas and, thus, are more susceptible to inundation than other areas. In addition, much of the marshland that provided a safety buffer against flooding has been filled and developed too. The case is made worse by the fact that these areas of fill are all subject to subsidence. San Francisco is sinking at −1.44 mm/year (Curry, 2018).

The threat of sea level rise is also a function of the settling of land. This is a particular problem in the Delta where small islands formed over 6000 years ago. At first, the island continues to rise as more growth occurs and dead vegetation collects on the island. That is partially offset by the loss of sediment that is blown or washed into the Bay and flows out into the Pacific Ocean. Later, wind causes erosion, and agricultural activity compacts the sediments and dead material to lower the inside of the island. The outer dike or levee prevents water from filling the island that is now lower than sea level. The space inside the island that is below sea level is called anthropogenic accommodation space. In 1900–2000, subsidence created approximately 2.5 billion m^3, and that number will rise to 3.0 billion m^3 by 2050 (Mount and Twiss, 2005). Continued subsidence from erosion and compaction and sea level rise increase the pressure on the levee and the risk of a breach (Fig. 7.1).

Numerous scientific studies have attempted to estimate the extent of the flooding. Estimates of inundation from sea level rise are complicated by land subsidence and uplift (Dangendorf et al., 2017). These are caused by multiple processes, including movement on geologic faults and mantle flow. In addition, consolidation of sediments, biological processes, and human activities can cause

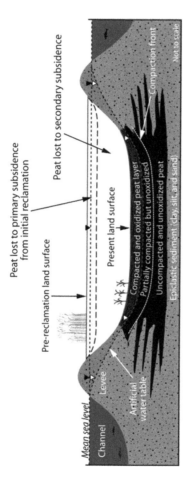

FIGURE 7.1 Subsidence in the Delta. Small islands formed in the Sacramento River Delta over 6000 years ago. Over time, plant growth caused the islands to grow higher as dead plant matter accumulated. It also created rich soil that farmers have taken advantage of for years. However, compaction due to agricultural activity and erosion now combine to lower the islands so that they are under sea level and must be protected by levees. The space inside the island that is below sea level is called anthropogenic accommodation space. In 1900–2000, subsidence created approximately 2.5 billion m^3, and that number will rise to 3.0 billion m^3 by 2050 (Mount and Twiss, 2005). Continued subsidence from erosion and compaction and sea level rise increase the pressure on the levee and the risk of a breach. Illustration is from USGS.

compaction or uplift. Shirzaei and Bűrgmann (2018) used radar interferometric measurement and global navigation data to determine the rates of subsidence around the Bay. Some areas were less than 2 mm/year, but others were more than 10 mm/year. These include parts of Treasure Island, San Francisco International Airport, and Foster City. They also created maps that estimate the 100-year inundation hazard. Those maps show that 125–429 km^2 may be flooded when subsidence is considered with sea level rise.

GROUNDWATER

The Central Valley of California has one of the world's largest aquifer systems. The agriculture industry in the Valley needs enormous quantities of water and far more than the traditional runoff of precipitation can supply. Thus, farmers there have been pumping out the ground water. The needs of agriculture, continuing population growth, and regular droughts have put a great deal of pressure on the ground water supply there. Ojha et al. (2018) correlated interferometric analysis of images acquired by synthetic aperture radar from satellites and GPS and groundwater data from 2007–2010 during a severe drought. They found that 2% of the acquire was permanently lost due to compaction. Their study provides the basis for future predictions.

BETWEEN LAND AND THE WATER

TIDAL AND MUD FLATS

Tidal flats surround much of the Bay. Those regions are the areas between the low and high tides that have a very low gradient. That is, they are flat. They might be mud or sand. They might also be referred to as salt flats since they are flooded with saltwater. Tidal mudflats are a key part of an estuary (van der Wegen et al., 2017). They alternate between being flooded and exposed to the air. They provide sediment for nearby salt marshes, food to birds and fish, and a home for many organisms. The shapes of tidal flat profiles have been related to such factors as the relative intensity of wave vs. tidal forcing, the supply and grain size of sediment, and local elevation of the flat with respect to mean sea level.

Although they have no obvious vegetation on the surface and their biodiversity is rather low, they are rich in life and important for recycling material from both terrestrial and marine sources. They dissipate wave action to protect the coastal regions and provide resting areas for resident and migratory birds. The flats can be divided into three parts: the supratidal zone is generally above the water, the intertidal zone is the area between the tides, and the subtidal is below the low tide mark and flooded most of the time.

The flats are dynamic. Sand or silts are constantly being deposited and eroded. The upper part of the intertidal zone tends to be composed of smaller particles and are called mudflats. The lower part tends to be sandy. Creeks running through the flats and the strength of the tides also affect the consistency of the flats. Water moving quickly carries heavier particles.

Mudflats have a higher organic load than sand flats. The material in mudflats is mostly silt and clay. They also differ in the amount of oxygen, and the availability of oxygen influences the type of organisms that live there. In the mudflats, the particles are smaller, and they trap organic material more effectively. The smaller particles have a greater surface area, which supports more bacteria and other microorganisms. The lower levels of oxygen result in mostly anerobic respiration that yields hydrogen sulfide, methane, and ammonia. Just below the surface (perhaps 1 cm), the silt is black and smells of sulfide. In sand flats, the grains are much larger, water penetrates it more easily, and the oxygen levels are higher. Sunlight also penetrates more deeply and, thus, allows for photosynthetic organisms.

South San Francisco Bay has an intermediate tidal range of up to 2.5 m (Bearman et al., 2010). The semidiurnal tides are mixed. Because the estuary is more narrow at the southern end, and this causes tidal waves to be reflected at the end. The Coriolis effect causes the tidal range to be a bit larger on the southwestern shore. Waves are due more to wind than to ocean swells. Erosion in the tidal flats is caused more by wind waves, and deposition is associated with tidal currents. Areas dominated by waves tend to have tidal flats composed of coarser sand.

In 2005, the intertidal flat area was a bit over 50 km^2. Half of the area was south of the Dumbarton Bridge and formed a low energy, higher tidal range. The width increases from 200 to 900 m as one moves from north to south. The sediments are primarily silts (~50 m), and they are mostly mud.

For the last several decades, the mudflats have been at equilibrium: depositions have essentially equaled erosion.

The tidal flats of the Bay Area are under increasing threats, and most are the result of human activities. The greatest threat is sea level rise. Estimates vary, but most are in the range of 60 cm in the next 100 years. Even that will have a huge effect on tidal flats. The change is far more rapid than has happened in thousands of years and will likely put much of the present day tidal flats under water.

Climate change is also causing heavier storms, and those storms are dumping greater than normal amounts of freshwater into the estuary. The increasing growth in the population of the Bay Area will add to the freshwater entering the Bay. The greater swings in salinity will be hard on many organisms. Pollution in the Bay could also increase. Human trading also brings invasive species.

SALT PONDS

The Bay contains saltwater, but the amount of salt varies according to the season and location. The coastal ocean is about 33–34 psu. Water near the Golden Gate is nearly as salty as the ocean, but deeper into the Bay and nearer the sources of freshwater, the salinity is lower. The South Bay surface water is about 93% seawater in dry seasons and about 50% seawater in wet seasons (Fig. 7.2).

FIGURE 7.2   Salt Ponds in the South Bay. Salt has been harvested from the Bay for many years. Seawater is allowed in to an intake pond and then moved through a series of evaporation ponds as the salinity continues to increase. As the salinity increases, the microorganisms in the water lend color that ranges from blue-green to magenta. The ponds support a great variety of shorebirds and waterfowl. Landsat photo from 2002.

Salt has been harvested from the Bay for centuries. Native Americans, such as the Ohlones, made salt from saltwater on a small scale. Like so much of the Bay Area's history, the demand for salt increased dramatically with the beginning of the Gold Rush. Without refrigeration, salt was needed to preserve food. Large areas of the South Bay were diked so that they could be flooded with Bay saltwater. Although they are not native, they have become embedded in the Bay Area environments and now support multiple species. The Bay is an important part of the Pacific Flyway for migratory waterfowl shorebirds. At least 30% of some groups use the salt ponds for staging and wintering. They are now a key part of that system.

The South Bay is idea for salt production. The bottom of the Bay and wetlands is a layer of clay, and the weather is generally mild. The process involves a series of evaporator ponds. Bay water is allowed to flood the first pond. The sun slowly evaporates the water and increases the salinity of the remaining water. The water is pumped from one pond to another until the water in the final most concentrated pond is moved to the treatment plant. The process takes about 5 years.

One of the most striking features of the salt ponds is their beautiful colors. At first, the water in the Bay is green or brownish color. As the salinity increases, the water becomes yellowish. Finally, it turns pink. That pink is due to the presence of microorganisms, such as halobacterium and algae. The populations rise and fall as the salinity increases. The organisms might thrive at one concentration, but then die off as it gets too salty. Others then take over. Early on algae begin to turn the water to a turquoise color. One common algal species is *Dunaliella salina*. These halophiles grow well in salty water. They contain the orange-red pigment beta-carotene. As the salt concentration increases and other algal species die off, the *D. salina* numbers increase, and the water becomes bright yellow and then orange from the beta-carotene. The beta-carotene protects the algae from damage due to UV irradiation. They also contain glycerol, which helps them withstand the osmotic pressures from the high salt. The brine shrimp in the salt ponds of San Francisco Bay are generally *Artemia franciscana*. They are common is other areas where saltwater evaporates, and they are an important food for many species. These crustaceans are very distantly related to the shrimp that we eat. To survive in the extremely salty conditions, they have evolved an efficient osmoregulatory system to preserve water and deal with the salt. As the salinity increases further, halobacteria cause the water to turn red. Despite their name, the halobacteria are not bacteria at all. They are archea, a distinct group of organisms.

Invertebrates in the salt ponds provide an excellent source of food for a number of waterbirds. The species and numbers of invertebrates vary with the salinity and depth. There are more at low salinity. Takekawa et al. (2009) examined common birds and their prey. American avocets liked seeds at low salinity and brine flies at mid-salinity. Western sandpipers took fewer prey types than avocets. Ruddy ducks are diving birds and foraged the lower depths for many different types of prey. No species seemed to depend much on brine shrimp. During their migrations, waterbirds (e.g., western sandpipers or *Calidris mauri*) tend to use the salt ponds heavily (Takekawa et al., 2006). This was especially true in the spring, possible because the populations of invertebrates are higher in the spring than other seasons. The ponds offer other advantages too. They facilitate take off, and they allow the birds to avoid predators and provide protected areas for roosting and foraging.

Cargill owns 12,000 acres of salt flats, from which they obtain 500,000 tons of salt each year. The company has turned over 40,000 acres to local conservation groups, and those groups are now working to restore the wetlands.

AIR AND CLIMATE

CLIMATE

San Francisco is famous for its fog. In summers, the air just over the low hills east of the San Francisco Bay heats up and rises. That results in cool moisture-laden Pacific air being pulled over the Bay. The water vapor condenses to form thick fog. In June 1579, Sir Francis Drake sailed north along the California coast looking for a safe harbor in which to repair his ship, the Golden Hind. Amazingly, he missed San Francisco Bay and landed further north at Point Reynes. Fog is thought to be the reason that Drake missed the Golden Gate when he was looking for a safe harbor. Or perhaps he passed by at night. The Golden Gate is barely a mile wide. It would have been easy to miss.

Coastal fog is actually a stratus or stratocumulus cloud that is very low or touches the ground. Fog forms because of the interaction of evaporation, aerosols, atmospheric pressure, air layering, temperature gradients, and topography. Fog provides needed moisture to plants and animals on the Bay Area coast, especially in the summer. However, the amount of fog has diminished in the Bay Area by 33% over the last century (Johnstone and Dawson, 2010). The USGS established the Pacific Coast Fog Survey in 2012 to collect basic data that would be useful to other researchers.

The climate of the Bay Area is referred to as a Mediterranean climate. It has warm dry summers (approximately April to October) and wet mild winters (approximately November to March). Weather patterns are influenced by the California current, which runs from British Columbia southward along the west coast to Baja Mexico. The current brings cooler temperatures than might be expected.

Weather in the Bay Area is further complicated by its diverse topography. That topography creates a number of micro-environments around the Bay. Fog and cooler air enter the Bay through gaps in the coastal range. One of these gaps, the Golden Gate gives San Francisco its cool foggy summers, even while the rest of the Bay Area might be quite warm. Other gaps to the south cool the South Bay cities, and others to the north cool the Marin headlands. Areas in the East Bay can be very warm. San Francisco is especially mild. Average temperatures are typically 45°F to 72°F. Temperatures in San Jose or in the East Bay, especially east of the East Bay hills, are much warmer. The average rainfall in the Bay Area is just under 24 inches per year.

The combination of hotter summers, drought, strong winds, and large amounts of fuel in forests and wild lands is resulting in more and larger wildfires in California. Wildfires in the Western United States have greatly increased in recent years.

Wildfires seem to be associated with more moisture in the growing season and less with the moisture during the fire season itself (Westerling et al., 2003). Westerling and Bryant (2008) modeled climate change and the risk of wildfires in four scenarios, including climate, hydrology, and topography. They confirmed that temperature during the fire season is critical as is moisture during the growing season. They also modeled the likelihood fires in suburban areas near major urban areas in the future.

The California wildfires in 2017 and 2018 were enormous, but wildfires in the state have been increasing in size and intensity since the 1970s. Williams et al. (2019) analyzed possible links between climate change caused by humans and increased wildfires in California. The strongest link that they found was that the drier air in the summer dries out the fuel supply to make it more prone to burn. Higher temperatures also contribute, especially in the fall. They also point out that the loss of moisture in the air that is caused by human activity is relatively small (about 10%) and that it is likely to double in the future. This is particularly troubling because the effect on fires is exponential.

The drought in the Southwestern United States continues. Williams et al. (2020) compared that drought to those from Medieval times to determine if climate change by human activity is involved. They examined tree rings over 1200 years to determine levels of soil moisture throughout the major droughts in 800 CE, late 1500s, and today. Clearly, natural factors were involved in all of them, but the recent drought also includes the effects of human activity in 31 climate models. They found that human activity and its effects on temperature, relative humidity, and precipitation accounted for about 47% of the recent drought severity. They conclude that human activity exacerbated a moderate drought to bring it to the severity of the worst droughts since 800 CE.

CLIMATE CHANGE

Climate change will significantly affect the Bay Area. In fact, several changes are already felt, according to California's Fourth Climate Change Assessment (Bedsworth et al., 2018). For example, between 1950 and 2005, the average annual maximum temperature increased by 1.7°F. The famous fog appears less often. The sea level has raised 20 cm in the last century. The 2012–2016 drought was the most severe in 1200 years.

The same report states that many of these changes will become worse in the next decades even if the nation initiates measures to reduce global warming. Temperatures will increase. Precipitation will become even more variable, swinging from drought to floods. The snow pack in the Sierras, which is so critical to the Bay Area's summer water supply, will be reduced by 20% in the next 20–30 years, and by 80% by the end of the century. The report predicts a sea level rise of 0.74–1.37 m by 2100. However, other research suggests that the increase might be nearly 3 m. Such an increase would result in the inundation of many areas around the Bay, especially in conjunction with stronger storms.

It is easy to imagine how such increases in temperature will affect coastal properties, water supplies, and energy demands. However, there will also be significant effects on plants and animals that will change the natural environments. For example, higher temperatures will stress redwoods, Douglas firs, and other evergreens. Chaparral will prosper. Wetlands will be lost to the rising sea levels. Hotter temperatures and less water will also stress many animals. Summer fires will increase in severity.

ATMOSPHERIC RIVERS

Water is a serious concern in California. The state seems to alternate between drought and floods. As the state has grown in population, competition for water among cities, farms, industry, and environmental concerns has also grown. Recently, scientists have begun to realize the role that atmospheric rivers play and have played in the water supplies in the state for hundreds of years. The atmospheric rivers result from low-level jets along the frontal edge of major winter cyclones in the eastern North Pacific. Satellite photographs of these storms show long trails of water vapor extending from California back to Hawaii. The storms are more than 2000 km long, a few hundred km wide, and in the lowest 2.5 km of the atmosphere.

Atmospheric rivers account for 20–50% of the state's rain for a year. Thus, the arrival of these storms or not is the difference between drought and plenty of water, and understanding them is critical for planning purposes. Dettinger et al. (2011) tracked a number of the factors (e.g., sea-surface temperatures and atmospheric conditions) in an attempt to predict atmospheric rivers. They suggest that use of these models could help California to better know how to use water resources. For the future, the models predict increasing variability in California precipitation. Decreases in levels of rain were not attributed to atmospheric rivers, but the increase in extreme rain levels is.

The Russian River just north of San Francisco has a long history of floods. For 8 years, Ralph et al. (2006) studied the effects of storms and atmospheric rivers on the river. They focused on one specific case of extreme weather that occurred on 16–18 February 2004. The storm matched the parameters that normally define an atmospheric river. The seven subsequent floods were consistent with the findings for the case study. However, not all atmospheric rivers result in floods. Other factors are important determinants. For example, very heavy rain after an extensive dry period did not result in a flood. Perhaps the soil had plenty of capacity to contain those storms.

BAY AIR

For the last 20–30 million years, the air above the San Francisco Bay has been unchanged. Before that time, now-dead volcanos belched out various gasses. Since then, periodic wild fires fouled the air with smoke. But there was little change otherwise. Air quality deteriorated with the arrival of humans. However, the smoke of Native Americans and even the early Spanish colonists contributed little pollution to Bay Area air. The arrival of large numbers of Europeans beginning in the mid-19th century began a great decline in air quality.

In fact, according to the "State of the Air 2020" report by the American Lung Association (2020), the Bay Area ranked third worst in the nation in 2020 in terms of short-term particle pollution and eight worst in the country for ozone pollution. The report implicates climate change as the responsible agent for the increased ozone levels. Unlike ozone levels, the levels of particulates are increasing.

In the last 15 years, wildfires have become as much a part of California summers as San Francisco fog. In 2007–2016, each year about 3000 fires are burned up to several hundred thousand acres. Fires can be characterized as dominated fuel or wind, and they are influenced by geographical distribution, fire history, source of ignition, seasonal timing, resources at risk, and responses. Keeley and Syphard (2019) showed that the Tubbs fire was much more damaging than previous ones because of the expansion of people and structures into previously undeveloped territory. High winds exacerbated the fires. The patterns of severe drought alternating with periods of high levels of precipitation have been occurring for hundreds of years. Precipitation in California is controlled by the strength and position of the North Pacific jet stream. Hot temperatures in the spring and summer result in very dry forests and scrublands. When these are combined with high winds, severe fires can result, such as those from 2017 to 2018 (Wahl et al., 2019).

The fires and the resulting smoke have significantly affected the air quality and the health and economy of the region. The recent spike in forest fires has a lot to do with the higher levels of particulates. The fires are also a function of climate change. During the years 2017 and 2018, fires raged throughout California and particular in the Bay Area. In fact, the air quality in San Francisco was the worst in recorded history due to the fires, and at one point in November 2018, the Bay Area had the worst air quality in the world. Many scientists expect that trend to continue as average temperatures soar. The quality of the Bay Area air will be seriously degraded. Of course, similar periods of poor air quality due to fires likely occurred in the past.

Gupta et al. (2018) used a network of low-cost air quality monitors and satellite data to better understand the distribution of microparticles. Microparticles (measure as $PM_{2.5}$ or particles with a diameter 2.5 m) are particularly hazardous. They found that the monitors were valuable for developing models of those distributions. Rooney et al. (2020) completed a similar study of the Camp Fire, one of the worst of the recent fires in Northern California in November 2018 and achieved similar results.

Methane and carbon monoxide are also important greenhouse gases. Fairley and Fischer (2015) measured their levels at 14 sites around the Bay Area and combined those results with global background levels. Methane levels were found to decrease slightly from 1990 to 2012. For technical reasons, they believe these are the lower estimates.

Air quality in the Bay Area has been greatly affected by human activity. Air pollution has a dramatic effect on human health, particularly lung and heart disease (Brunekreef and Holgate, 2002). In the Bay Area, for example, Kim et al. (2004) examined the correlation between traffic pollutants and local air quality. They measured pollutants (e.g., particulates, black carbon, nitrogen oxides, and nitrogen dioxide) at 10 school sites around the Bay Area. They found an association of respiratory symptoms (e.g., asthma) and pollutants.

Climate has already begun to change air quality in the Bay Area. These include warmer average temperatures, diminished rainfall, and rising sea levels. With new regulations, the air has improved greatly in the last decades. However, climate change will put those gains at risk.

California is a large state and an important contributor to greenhouse gases in the United States. About 7% of total US greenhouse gases come from California. Transportation is the largest single contributor. The state has used legislation to attempt to reduce levels of these pollutants in 2050 to less than their levels in 1990.

Those efforts have been effective in reducing levels of greenhouse gases and also in improving public health. Median daily walking and bicycling increased from 4 to 22 minutes lowered cardiovascular disease and diabetes, and decreased greenhouse gases (Maizlish et al., 2013).

California's climate is changing due to global warming. Recent years have been much drier on average and the state has been experiencing a significant drought. The normal "two-season" year has been disrupted. Higher temperatures for longer periods have greatly increased the risk of wildland fires and extended the fire season for additional months. Cloern et al. (2011) predict that these trends will continue throughout the 21st century and that the Bay Area climate will see dramatic changes.

REFERENCES

Aagaard B.T., Blair J.L., Boatwright J., Garcia S.H., Harris R.A., Michael A.J., Schwartz D.P., DiLeo J.S. (2014) Earthquake Outlook for the San Francisco Bay Region 2014–2043. *USGS*, https://pubs.usgs.gov/fs/2016/3020/fs20163020.pdf, accessed December 6, 2020.

Bedsworth L., Cayan D., Franco G., Fisher L., Ziaja S., (California Governor's Office of Planning and Research, Scripps Institution of Oceanography, California Energy Commission, California Public Utilities Commission) (2018) Statewide Summary Report. *California's Fourth Climate Change Assessment*. Publication number: SUMCCCA4-2018-013.

American Lung Association (2020) State of the Air 2020. https://www.stateoftheair.org/city-rankings/most-polluted-cities.html   https://www.lung.org/assets/documents/healthy-air/state-of-the-air/sota-2018-full.pdf, accessed December 7, 2020.

Auta H.S., Emenike C.U., Fauziah S.H. (2017) Distribution and importance of micro-plastics in the marine environment: A review of the sources, fate, effects, and potential solutions. *Environment International* 102: 165–176.

Ayres D.R., Smith D.L., Zaremba K., Klohr S., Strong D.R. (2004) Spread of exotic cordgrasses and hybrids (*Spartina* sp.) in the tidal marshes of San Francisco Bay, California, USA. *Biological Invasions* 6: 221–231.

Barnard P.L., Schoellhamer D.H., Jaffe B.E., McKee L.J. (2013) Sediment transport in the San Francisco Bay Coastal System: an overview. *Marine Geology* 345: 3–17.

Bearman J.A., Friedrichs C.T., Jaffe B.E., Foxgrover A.D. (2010) Spatial trends in tidal flat shape and associated environmental parameters in South San Francisco Bay. *Journal of Coastal Research* 26: 342–349.

Berg G.M., Driscoll S., Hayashi K., Kudela R. (2019) Effects of nitrogen source, con-centration, and irradiance on growth rates of two diatoms endemic to northern San Francisco Bay. *Aquatic Biology* 28: 33–43.

Brew D.S., Williams P.B. (2010) Predicting the impact of large-scale tidal wetland re-storation on morphodynamics and habitat evolution in South San Francisco Bay, California. *Journal of Coastal Research* 26: 912–924.

Brunekreef B., Holgate S.T. (2002) Air pollution and health. *The Lancet* 360: 1233–1242.

Cappiella K., Malzone C., Smith R., Jaffe B. (1999) Sedimentation and bathymetry changes in Suisun Bay: 1867–1990. *US Geologic Survey, Open-File Report* 99–563.

Cayan D.R., Polade S.D., Kalansky J., Ralph F.M. (2019) Precipitation regime change in Western North America: the role of atmospheric rivers. *Scientific Reports* 9: 9944.

Charles J. (2014) Fog's fingerprint on coastal ecology. *Estuary News* http://www.sfestuary. org/wp-content/uploads/2014/03/CALCC-Mar2014-v7x-web.pdf, accessed December 7, 2020.

Cloern J.E., Jassby A.D. (2012) Drivers of change in estuarine-coastal ecosystems: dis-coveries from four decades of study in San Francisco Bay. *Reviews of Geophysics* 50: RG4001 doi:10.1029/2012RG000397.

Cloern J.E., Jassby A.D., Thompson J.K., Hieb K.A. (2007) A cold phase of the East Pacific triggers new phytoplankton blooms in San Francisco Bay. *Proceedings of the National. Academy of Sciences* 104: 18561–18565.

Cloern J.E., Knowles N., Brown L.R., Cayan D., Dettinger M.D., Morgan T.L., et al. (2011) Projected evolution of California's San Francisco Bay-Delta-River system in a century of climate change. *PLoS One* 6(9): e24465.

Cohen-Waeber J., Sitar N., Bürgmann R. (2013) GPS Instrumentation and Remote Sensing Study of Slow Moving Landslides in the Eastern San Francisco Bay Hills, California, USA. *Proceedings of the 18th International Conference on Soil Mechanics and Geotechnical Engineering*, Paris, pp. 2169–2172.

Cordeira J.M., Stock J., Dettinger M.D., Young A.M., Kalansky J.F., Ralph F.M. (2019) A 142-year climatology of Northern California landslides and atmospheric rivers. *Bulletin of the American Meteorological Society* 100: 1499–1509.

Curry J. (2018) Sea Level and Climate Change. *Climate Forecast Applications Network,* https://www.junkscience.com/wp-content/uploads/2019/02/special-report-sea-level-ris e3.pdf, accessed December 7, 2020.

Dallas K.L., Barnard P.L. (2011) Anthropogenic influences on shoreline and nearshore evolution in the San Francisco Bay coastal system. *Estuarine, Coastal and Shelf Science* 92: 195–204.

Dangendorf S., Marcos M., Woppelman G., Conrad C.P., Frederikse T., Riva R. (2017) Reassessment of 20th century global sea level rise. *Proceedings of the National Academy of Sciences USA* 114: 5946–5951.

De Falso F., Di Pace E., Cocca M., Avella M. (2019) The contribution of washing processes of synthetic clothes to microplastic pollution. *Scientific Reports* 9: 6633.

Dettinger M.D., Ralph F.M., Das T., Neiman P.J., Cayan D.R. (2011) Atmospheric rivers, floods and the water resources of California. *Water* 3: 445–478.

Donovan P.M., Blum J.D., Yee D., Gehrke G.E. (2013) An isotopic record of mercury in San Francisco Bay sediment. *Chemical Geology* 349–350: 87–98.

Eagles-Smith C.A., Wiener J.G., Eckley C.S., Willacker Jr J.J., Evers D.C., Marvin-DiPasquale M.C., Obrist D., Fleck J.A., Aiken G.R., Lepak J.M., Jackson A.K., Stewart A.R., Webster J., Davis J.A., Alpers C.N., Ackerman J.T. (2016) Mercury in western North America—A synthesis of environmental contamination, fluxes, bioaccumulation and risk to fish and wildlife. *Science of the Total Environment* 568: 1213–1226. http://dx.doi.org/10.1016/j.scitotenv.2016.05.094.

Elder W.P. (2013) Bedrock geology of the San Francisco Bay Area: A local sediment source for bay and coastal systems. *Marine Geology* 345: 18–30.

Fairley D., Fischer M.L. (2015) Top-down methane emissions estimates for the San Francisco Bay Area from 1990 to 2012. *Atmospheric Environment* 107: 9–15.

EPA (2020) Hunters Point Naval Shipyard, San Francisco, CA. https://cumulis.epa.gov/supercpad/cursites/csitinfo.cfm?id=0902722, accessed December 7, 2020.

Ferretti A., Novali F., Brűgmann R., Hilley G., Prati C. (2004) InSAR permanent scatterer analysis reveals ups and downs in San Francisco Bay Area. *Eos* 85: 317–324.

Finnegan N.J., Broudy K.N., Nereson A.L., Roering J.J., Handwerger A.L., Gennett G. (2019) River channel width controls blocking by slow-moving landslides in California's Franciscan mélange. *Earth Surface Dynamics* 7: 879–894.

Foxgrover A.C., Higgins S.A., Ingraca M.K., Jaffe B.E., Smith R.E. (2004) Deposition, Erosion, and Bathymetric Change in South San Francisco Bay: 1858–1983. US Geologic Survey Open-File Report 2004-1192. https://pubs.er.usgs.gov/publication/ofr20041192, accessed December 7, 2020.

Gershunov A., Shulgina T., Clemesha R.E.S., Guirguis K., Pierce D.W., Dettinger M.D., Lavers D.A., Dijkstra Y.M., Schuttelaars H.M., Burchard H. (2017) Generation of exchange flows in estuaries by tidal and gravitational eddy viscosity-shear covariance (ESCO). *Journal of Geophysical Research: Oceans* 122: 4217–4237.

Gibbs R.J. (1977) Transport phases of transition metals in the Amazon and Yukon Rivers. *Geological Society of America Bulletin* 88: 829–843.

Greenfield B.K., Davis J.A., Fairey R.I., Roberts C., Crane D., Ichikawa G. (2005) Seasonal, interannual, and long-term variation in sport fish contamination, San Francisco Bay. *Science of the Total Environment* 336: 25–43.

Greenfield B.K., Jahn A. (2010) Mercury in San Francisco Bay forage fish. *Environmental Pollution* 158: 2716–2724.

Gupta P., Doraiswamy P., Levy R., Pikelnaya O., Maibach J., Feenstra B., Polidori A., Kiros F., Mills K.C. (2018) Impact of California fires on local and regional air quality: the role of a low-cost sensor network and satellite observations. *GeoHealth* 2: 172–181.

Hart J., Sanger D. (2003) *San Francisco Bay. Portrait of an Estuary*. University of California Press, Berkeley.

Higgins S.A., Jaffe B.E., Fuller C.C. (2007) Reconstructing sediment age profiles from historical bathymetry changes in San Pablo Bay, California. *Estuarine, Coastal and Shelf Science* 73: 165–174.

Ingram B.L., Ingle J.C., Conrad M.E. (1996) A 2000 yr record of Sacramento–San Joaquin river inflow to San Francisco Bay estuary, California. *Geology* 24: 331–334.

Jaffe B.E., Smith R.E., Foxgrover A.C. (2007) Anthropogenic influence on sedimentation and intertidal mudflat change in San Pablo Bay, California: 1856–1983. *Estuarine, Coastal and Shelf Science* 73: 175e187.

Johnson S., Gordon L. (2015) New maps reveal seafloor off San Francisco area. *Sound Waves. US Geologic Survey.* https://soundwaves.usgs.gov/2015/06/pubs.html, accessed December 7, 2020.

Johnstone J.A., Dawson T.E. (2010) Climatic context and ecological implications of summer fog decline in the coast redwood region. *Proceedings of the National Academy of Sciences* 107: 4533–4538.

Keeley J.E., Syphard A.D. (2019) Twenty-first century California, USA, wildfires: Fuel-dominated vs. wind-dominated fires. *Fire Ecology* 15: 24. https://doi.org/10.1186/s42408-019-0041-0.

Kim J.J., Smorodinsky S., Lipsett M., Singer B.C., Hodgson A.T., Ostro B. (2004) Traffic-related air pollution near busy roads. The East Bay Children's Respiratory Health Study. *American Journal of Respiratory and Critical Care Medicine* 170: 520–526.

Kimmerer W.J. (2002) Physical, biological, and management responses to variable freshwater flow into the San Francisco Estuary. *Estuaries* 25: 1275–1290.

Kjelland M.E., Woodley C.M., Swannack T.M., Smith D.L. (2015) A review of the potential effects of suspended sediment on fishes: Potential dredging-related physiological, behavioral, and transgenerational implications. *Environment Systems and Decisions* 35: 334–350.

Lacroix P., Handwerger A.L., Bièvre G. (2020) Life and death of slow-moving landslides. *Nature Reviews Earth and Environment* 1: 404–419.

Maizlish N., Woodcock J., Co S., Ostro B., Fanai A., Fairley D. (2013) Health cobenefits and transportation-related reductions in greenhouse gas emissions in the San Francisco Bay Area. *American Journal of Public Health* 103: 703–709.

Mason S.A., Garneau D., Sutton R., Chu Y., Ehmann K., Barnes J., Fink P., Papazissimos D., Roger D.L. (2016) Microplastic pollution is widely detected in US municipal wastewater treatment plant effluent. *Environmental Pollution* 218: 1045–1054.

Masura J., Baker J., Foster G., Arthur C., Herring C. (2015) Laboratory methods for the analysis of microplastics in the marine environment: recommendations for quantifying synthetic particles in waters and sediments. NOAA Technical Memorandum NOS-OR&R-48.

MacVean L.J., Lewis L.S., Trowbridge P., Hobbs J.A., Senn D.B. (2018) Dissolved oxygen in South San Francisco Bay: Variability, important processes, and implications for understanding fish habitat. Technical Report. San Francisco Estuary Institute, Richmond, CA.

McKee L.J., Ganju N.K., Schoellhamer D.H. (2006) Estimates of suspended sediment entering San Francisco Bay from the Sacramento and San Joaquin Delta, San Francisco Bay, California. *Journal of Hydrology* 323: 335–352.

Moftakhari H.R., Jay D.A., Talke S.A., Schoellhamer D.H. (2015) Estimation of historic flows and sediment loads to San Francisco Bay, 1849–2011. *Journal of Hydrology* 529: 1247–1261.

Monismith S.G., Kimmerer W., Burau J.R., Stacey M.T. (2002) Structure and flow-induced variability of the subtidal salinity field in Northern San Francisco Bay. *Journal of Physical Oceanography* 32: 3003–3019.

Mount J., Twiss R. (2005) Subsidence, sea level rise, seismicity in the Sacramento-San Joaquin Delta. *San Francisco Estuary and Watershed Science* 3(1). http://repositories.cdlib.org/jmie/sfews/vol3/iss1/art5.

Nichols F.H., Cloern J.E., Luoma S.N., Peterson D.H. (1986) The modification of an estuary. *Science* 231: 567–573.

Odum E.P. (1969) The strategy of ecosystem development. *Science* 164: 262–270.

Ojha C., Shirzaei M., Werth S., Argus D.F., Farr T.G. (2018) Sustained groundwater loss in California's Central Valley exacerbated by intense drought periods. *Water Resources Research* 54: 4449–4460.

Oros D.R., Ross J.R.M., Spies R.B. (2007) Polycyclic aromatic hydrocarbon (PAH) contamination in San Francisco Bay: a 10-year retrospective of monitoring in an urbanized estuary. *Environmental Research* 105: 101–118.

Ralph F.M., Neiman P.J., Wick G.A., Gutman S., Dettinger M., Cayan D., White A.B. (2006) Flooding on California's Russian River—Role of atmospheric rivers. *Geophysical Research Letters* 33: 5.

Respaut R., Levinson R. (2019) A California Naval base shutters, and contamination lingers decades later. *Reuters.* https://www.reuters.com/investigates/special-report/usa-military-legacy/.

Rooney B., Wang Y., Jiang J.H., Zhao B., Zeng Z.-C., Seinfeld J.H. (2020) Air quality impact of the Northern California camp fire of November 2018. *Atmospheric Chemistry and Physics* 20: 14597–14616. https://doi.org/10.5194/acp-2020-541.

Rudnick D.A., Kieb K., Grimmer K.F., Resh V.H. (2003) Patterns and processes of biological invasion: the Chinese mitten crab in San Francisco Bay. *Basic and Applied Ecology* 4: 249–262.

Schraga T.S., Cloern J.E. (2017) Water quality measurements in San Francisco Bay by the U.S. Geological Survey, 1969–2015, *Scientific Data* 4: 170098. https://www.nature.com/articles/sdata201798, accessed December 7, 2020.

Schmidt D.A., Bűrgmann R. (2003) Time-dependent land uplift and subsidence in the Santa Clara valley, California, from a large interferometric synthetic aperture radar data set. *Journal of Geophysical Research* 108: 2416–2428.

Schoelhamer D.H., Mumley T.E., Leatherbarrow J. (2007) Suspended sediment and sediment-associated contaminants in San Francisco Bay. *Environmental Research* 105: 119–131. https://www.sciencedirect.com/science/article/pii/S0013935107000424.

Semler M.O., Blanchard R.L. (1989) Radiological survey of the Mare Island Naval Shipyard, Alameda Naval Air Station, and Hunters Point Shipyard. *National Technical Reports Library.* https://ntrl.ntis.gov/NTRL/dashboard/searchResults/titleDetail/PB89210637.xhtml, accessed December 7, 2020.

Shirzaei M., Bűrgmann R. (2018) Global climate change and local land subsidence exacerbate inundation risk to the San Francisco Bay Area. *Science Advances* 4: eaap9234.

Singer M.B., Aalto R., James L.A., Kilham N.E., Higson J.L., Ghoshal S. (2013) Enduring legacy of a toxic fan via episodic redistribution of California gold mining debris. *PNAS* 110: 18436–18441 https://doi.org/10.1073/pnas.1302295110.

Stralberg D., Brennan M., Callaway J.C., Wood J.K., Schile L.M., Jongsomjit D., Parker V.T., Crooks S. (2011) Evaluating tidal marsh sustainability in the face of sea-level rise: a hybrid modeling approach applied to San Francisco Bay. *PLoS One Online Journal* 6(11): e27388. https://doi.org/10.1371/journal.pone.0027388.

Sutton R., Mason S.A., Stanek S.K., Willis-Norton E., Wren I.F., Box C. (2016) Microplastic contamination in the San Francisco Bay, California, USA. *Marine Pollution Bulletin* 109: 230–235.

Takekawa J.Y., Miles A.K., Schoelhamer D.H., Atherarn N.D., Saiki M.K., Duffy W.D., Kleinschmidt S., Schellenbarger G.G., Jannusch C.A. (2006) Trophic structure and avian communities across a salinity gradient in evaporation ponds of the San Francisco Bay estuary. *Hydrobiologia* 567: 307–327.

Takekawa J.Y., Miles A.K., Tsao-Melcer D.C., Schoelhamer D.H., Fregien S., Athearn N.D. (2009) Dietary flexibility in three representative waterbirds across salinity and depth gradients in salt ponds of San Francisco Bay. *Hydrobiologia* (2009) 626: 155–168.

van der Wegen M., Jaffe B., Foxgrover A., Roelvink D. (2017) Mudflat morphodynamics and the impact of sea level rise in South San Francisco Bay. *Estuaries and Coasts* 40: 37–49.

Voosen P. (2020) A muddy legacy. *Science* 369: 898–901.

Vroom J., Van der Wegen M., Martyr-Koller R.C., Lucas L.V. (2017) What determines water temperature dynamics in the San Francisco Bay-Delta system? *Water Resources Research* 53: 9901–9921. https://doi.org/10.1002/2016WR020062.

Wahl E.R., Zorita E., Trouet V., Taylor A.H. (2019) Jet stream dynamics, hydroclimate, and fire in California from 1600 CE to present. *Proceedings of National Academy of Sciences USA* 116: 5393–5398.

Westerling A.L., Brown T.J., Gershunov A., Cayan D.R., Dettinger M.D. (2003) Climate and wildfire in the Western United States. *Bulletin of American Meteorological Society* 84: 595–604.

Westerling A.L., Bryant B.P. (2008) Climate change and wildfire in California. *Climatic Change* 87(Suppl 1): S231–S249.

Williams A.P., Abatzoglou J.T., Gershunov A., Guzman-Morales J., Bishop D.A., Balch J.K., Lettenmaier D.P. (2019) Observed impacts of anthropogenic climate change on wildfire in California. *Earth's Future* 7: 892–910.

Williams A.P., Cook E.R., Smerdon J.E., Cook B.I., Abatzoglou J.T., Bolles K., Baek S.H., Badger A.M., Livneh B. (2020) Large contribution from anthropogenic warming to an emerging North American megadrought. *Science* 368: 314–318.

8 Biology of the Bay

The San Francisco Bay Area is in a Temperate-Mediterranean Climate Zone also known as the dry summer climate. The climate is characterized by rainy winters and dry summers, with less than 40 mm of precipitation for at least three summer months. During the summer (June through August) it gives warm inland areas (>25°C) and cooler coastal regions (<25°C) followed by mild winters (December through February) on the coasts with occasional frosts inland (Federal Records, 2020). The flora and fauna are typical of such a climate zone and many of the species endemic to the region are also to be found in similar climate zones in Eurasia (Jiang et al., 2019).

Regarding a description of the flora and fauna of the Bay Area, it is not our intention to provide an encyclopedic list of all the resident species. Nor should this section be relied upon as a definitive accounting of the flora and fauna of the Bay Area. Rather, we will give a few chosen examples of how the environment has affected (and sometimes changed) the evolution of various organisms.

ANIMALS

VERTEBRATES

This section includes descriptions of all the extant animals that are classified as chordates, implying they have had at least a dorsal notochord and/or a nerve cord during ontogeny (embryonic growth and development) (Rychel et al., 2006).

As described earlier, the megafauna of the Upper Pleistocene period (129–11.65 cal kya) were predominantly mammoths, mastodons, musk ox, woolly rhinoceros, giant sloths, horses, camels, the monstrous short-face bear, a number of large saber-toothed cats, and the dire wolf, in addition to the ancestors of those that survived into our present time, the Holocene (11.65 cal kya to the present).

We will introduce the description of the fauna of the Bay Area by the way of Order and the Family. We will only include those animals which have in past or currently occupy ecological niches in the Bay Area; we may also mention if a species that is considered endemic (a species confined to a specific ecological or geographical area) to the Bay Area is also found elsewhere, although this is very rare (e.g., salamanders of the genus *Ensatina*, as described later). We will describe in classic trophic terms, from the top predators through the herbivores and those smaller animals that depend upon both groups.

Mammals

Predators

The first are the Carnivora, the carnivores, which in turn comprise the following Families: Ursidae (bears), Canidae (dogs and their relatives), Mustelidae (weasels, martens, otters, badgers, skunks), Procyonidae (raccoons), Otariidae (eared seals), Phocidae (earless seals), and Felidae (all cats) (Flynn et al., 2005; Wilson and Mittermeier, 2009). The large or the largest predators in an ecosystem are usually referred to by biologists and ecologists as the "keystone predator." In this context, the term "keystone" refers to the influential niche position of the predator within the ecosystem and which contributes to how the ecosystem is maintained in balance (Beschta and Ripple, 2009; Wallach et al., 2010; Ripple et al., 2014). This is referred to as a top-down trophic cascade (Leopold 1949; Hairston et al., 1960; Oksanen et al., 1981). Briefly, a keystone predator maintains an ecosystem in balance by predation upon herbivores that themselves feed upon vegetation that is critical to maintaining the structure of the ecosystem; for example, by securing the banks of a stream or river so that flooding and sedimentary buildup is controlled. If the keystone predator is removed, such as by human control measures or by disease, the prey herbivores multiply uncontrollably, devour more vegetation, and with less vegetation to secure the banks of the stream or river, the fluvial flow is compromised, leading to flow rate changes, build-up of silt and eventual change of the topography and geomorphology (Beschta and Ripple, 2009).

Although the large keystone predators (grizzly bear, *Ursus arctos californicus*) and grey wolves, (*Canis lupus*) had been hunted to extinction by the early 1900s (Miller and Waits, 2006) there still remain a number of keystone predators in the region. At one time, keystone predators were considered to be detrimental to livestock and danger to people, and so during most of the 18th, 19th, and 20th centuries they were allowed to be hunted at will, frequently with bounties for each claimed corpse; in California, hunting of mountain lions was not banned until 1990 under California Proposition 117 (California Fish and Game Code Div. 3, Ch. 9, Art. 2, 2785-2799.6). During the late 1980s and onwards, ecologists and other biologists determined that removing these keystone predators from the ecosystem led to significant damage to the environment (Edvenson, 1994; Beschta and Ripple, 2009).

Scientists now consider that the grizzly bear is really just a larger brown bear (also *Ursus arctos*) that migrated from Eurasia after the last Ice Age about 14 kya (1,000 years ago) as they have no significant mitochondrial DNA differences (Cronin et al., 1991).

The remaining and extant large predators include the American black bear (*Ursus americanus californiensis* in the Sierra Nevada and *U. a. altifrontalis* in the north coast and Cascades), mountain lion (*Puma concolor*), also known as the puma, cougar, or catamount (probably an Anglicization of the Spanish for mountain lion, "gato monte"), and the coyote (*Canis latrans*).

Black bears are found in the coastal ranges north of San Francisco, but the populations are small (CDFW, nd-a). In the San Francisco Bay Area, cougars and coyotes have become more tolerant to humans and encounters within municipal boundaries are often reported in the local news (Rodriguez, 2020; Presidio Trust, 2020). Residents at the periphery of the urban and suburban communities may often hear the squealing and yipping of coyote pups greeting their parents return after a successful night's hunt in the mid-summer.

On a side note, it is of interest that some of the oldest fossils of the coyote have been found in Irvington, a district of the City of Fremont (southern Alameda County), dated at around 0.73 Mya (million years ago) (Edford et al., 2009).

Another more-rarely spotted predator is the red fox (*Vulpes vulpes* or *Vulpes vulpes fulva*) and these too have become habituated to urban and suburban life. Interestingly, these red foxes are not part of the fox clades inhabiting the southern (or montane) refugium (subalpine parklands and alpine meadows of the Rocky Mountains, the Cascade Range, and Sierra Nevada), but are descended from the North American eastern red fox, introduced to the lower part of California during the 1870s for the fur trade and fox hunting. Some of these have mated with the Sacramento Valley red fox (*v. v. fulva patwin*) and there exists a narrow hybrid zone west of the Sacramento River and south of US Interstate 80 (Solano County and northern Contra Costa County) (Jurek, 1992; Sacks et al., 2011).

The mountain lion (*Puma concolor*), also known as the puma or cougar, is to be found throughout the Bay Area grasslands and woodlands, particularly in the North Bay, East Bay, and South Bay hills, and the coastal ranges farther south. Although frequently encountered in the Bay Area, there have been moves recently to temporarily protect pumas on the Central Coast through to Southern California under the Endangered Species Act (ESA) (Sahagún, 2020). Their natural prey are deer, coyotes, galliform birds (such as turkey; *Meleagris gallopavo*, the California quail; *Callipepla californica*, as well as rodents and insects) (Bay Area Puma Project, nd).

The bobcat (*Lynx rufus*) is up to two times larger than the domestic cat, standing 38 cm (15"), and weighs about 9 kg (20 lb; females) and between 7 and 13.5 kg (16–30 lb; males) (CDFW, nd-b). They are present throughout the Bay Area but as they are generally nocturnal, are observed infrequently. During the day, to avoid larger predators (and humans) they rest in what is termed as "scrape," which is a shallow pit scraped out with their hind limbs. The bobcat is most likely descended from the Eurasian lynx, having migrated across the Bering Land Bridge (Beringia) about 2.6 Mya (Johnson and O'Brien, 1997; Pecon-Slattery and O'Brien, 1998) but then became isolated during the subsequent Ice Ages.

The Musteloidea are a superfamily of the Order Carnivora and to which belong many of the small predators common in the Bay Area. They include the Procyonidae (raccoons and their allies, named for their original classification as "pre-dogs"), the Mustelidae (otters, badgers, weasels, and their allies, named for the musk (scent/must) gland used to mark territory), the Mephetidae (skunks), and the Ailuridae (the red panda).

The common raccoon (*Procyon lotor*) is the largest and the most prevalent member of the procyonid family in North America, but unlike most other carnivores, it is an opportunistic omnivore (Wozencraft, 2005). They are defined as a mesopredator, which are characterized by the rapid population growth of intermediate-sized predators in the absence of larger top predators (Martin, 2011).

The name "raccoon" is derived from the Algonquian word *aroughcoune*, meaning "he who scratches with his hands." The Algonquian nation (who call themselves *Omàmiwininiwak* and *Anicinàpe*) is in northeastern North America (present-day Québec and eastern Ontario) and who had traded raccoon furs with the French during the 16th–18th centuries (Litalien, 2004; Poulter, 2010). Raccoons are closely related to the ringtail (*Bassariscus sp.*), found in the dryer parts of Oregon, Eastern California, and extending to the southwestern United States, the olingos (*Bassaricyon sp.*), and the coati (*Nasua* and *Nasuella spp.*) and are more distantly related to the kinkajou (*Potos flavus*); however, the latter species are only found in Central and South America (Wozencraft, 2005). It is largely nocturnal, digging up worms and small invertebrates, and small family groups are often seen wandering the suburbs of the Bay Area (Martin, 2011). Whilst they may hibernate during the winter in other parts of North America, the Bay Area climate is sufficiently mild that raccoons active throughout the year. Although other procyonids are generally form social groups, raccoons differ in that, other than sow-kit groups, they tend to remain solitary and may engage in fights during encounters with other raccoons (Barrat, 2013).

Members of the family Mustelidae including badgers, otters (both sea and river subspecies), weasels, mink, and martens, are common throughout the Bay Area (Martin, 2011); these are also termed mesopredators and we will now describe them in more detail.

The American badger (*Taxidea taxus jeffersoni*), is not as common in much of the Bay Area, particularly in areas developed for human activity. They are, however, more abundant in the northern counties in protected and private lands, such as in Sonoma County and the Point Reyes National Seashore. It is among the largest of the mustelids in North America, males reaching between 60 and 75 cm (23.5″ and 29.5″) in length and up to 8.6 kg (19 lb) in weight; the females, on the other hand, are general smaller, at between 6.3 and 7.2 kg (14–16 lb) (Feldhamer et al., 2003). This is termed "sexual dimorphism" and is to be found throughout the mammalian class (Klymkowsky et al., 2016). It is an aggressive animal, although its young can be the target of the golden eagle (*Aquila chrysaetos*), mountain lion, coyote, or bobcat. It is primarily a fossorial carnivore (the word is derived from Latin, "fossa," a ditch), meaning that it hunts mammals and birds that usually reside in burrows or underground tunnel systems. Its prey in the Bay Area is the California ground squirrel (*Otospermophilus beecheyi*), the California vole (*Microtus californicus*), pocket gophers (*Thomomys bottae*), moles (*Scapanus latimanus*), and snakes, including rattlesnakes (*Crotalus spp.*).

The sea otter (*Enhydra lutris*) is native to the west coast of North America and is distinct from the Eurasian otter (*Lutra lutra*). It is the heaviest of the mustelids and unlike the others, has no scent gland (Kenyon, 1969). Sea otters make their home amongst the giant kelp forests growing in the shallow waters of the coast (see above) and its shelter provides much of the food for them, mainly echinoderms, such as sea urchins, crustaceans, mollusks, and fish, and it is a keystone species. A close relative, the North American river otter (*Lontra canadensis*), inhabits fresh water, and is often seen in the Sacramento Delta. Its prey differs from that of the sea otter in that it principally consumes fish, amphibians, mollusks, and crustaceans (Grenfell, 1974; Larsen, 1984; Melquist and Dronkert, 1987).

Other common mustelids resident in the Bay Area are the long-tailed weasel (*Mustela frenata*), American mink (*Neovison vison*), and North American martens (*Martes americana*), the latter is generally found in the more mountainous parts of the Bay Area.

The striped skunk (*Mephitis mephitis*) was once included within the mustelids, but is now classified in its own family, Mephetidae, and like the raccoon, is an omnivore (Dragoo and Honeycutt, 1997). Like the raccoon, it is primarily nocturnal although it can be easily detected during the daylight hours by its pungent smell, originating in its anal scent glands, situated ventral to its tail.

A visitor to the commercial piers on the northern edge of San Francisco is unlikely to miss the large marine predator, the California sea lion (*Zalophus californianus*). It is in the family Otariidae, the eared-seals, having external ear flaps, and which differ from true seals, which lack external ears. They are also distinguished from the true seal in that their front flippers are long and strong enough to hold themselves upright and they have rear flippers which are used for propulsion on land, facing forwards. There is also another species of sea lion common along the Bay Area coast, the Steller sea lion (*Eumetopias jubatus*) but it is much larger and lighter in color than the California sea lion. They have a breeding colony on the Island of Año Neuvo, which lies approximately 80 km (50 miles) south of San Francisco.

Another visitor to the Bay Area is the northern fur seal (*Callorhinus ursinus*), also in the family Otariidae, but these rarely come ashore on the mainland and are mainly found on the Farallon Islands, where they established a breeding colony (rookery) in the mid-1990s or out at sea, hunting. They are a protected species under the Marine Mammal Protection Act of 1972.

True seals, on the other hand, are represented in the Bay Area by the Pacific harbor seal (*Phoca vitulina*). They are more specialized for aquatic life than are sea lions, having lost the ability to ambulate using their flippers on land.

The northern elephant seal (*Mirounga angustirostris*) has become a more common sight during the past 30 to 40 years in the Bay Area than in the past (Abadía-Cardoso et al., 2017) when they became protected under the Marine Mammal Protection Act. Since then, their numbers have increased spectacularly and have probably reached the population size before they became targets for their oil. They have a breeding colony on the Island of Año Neuvo and the adjacent shoreline; they are now part of tours guided by Park Rangers at Año Neuvo State Park during the winter months. There is a much smaller breeding colony at Drakes Beach and the adjacent south-facing beaches of the headlands in the Point Reyes National Seashore, 63 km (39 miles) north of San Francisco (Elephant Seal Overlook).

Whales

In the last century, whales were hunted almost to extinction (Rice, 1998). By the 1940s, there was sufficient concern amongst the scientific and many western industrial communities that a moratorium on hunting should be enacted. The International Whaling Commission enacted such a moratorium in 1946 and 1982, although some nations and aboriginal communities were permitted to continue at much reduced levels ("The Economist" article 2012). Since those times, baleen whale populations have increased significantly and residents of the Bay Area are able to observe up close migrating grey whales (Eschrichtiidae), humpbacks and blue whales (Balaenopteridae), and sperm whales (Physeteridae) at many times of the year.

In addition to the magnificent immense baleen whales, smaller toothed whales, such as dolphins (Delphinidae) and porpoises (Phocoenidae) are also frequently observed coursing back and forth along the coast, preying upon schools of fish, usually sardines or anchovy.

Deer

The vast majority of large prey mammals familiar to residents of the Bay Area are deer, predominantly the Columbian black-tailed deer (*Odocoileus hemionus columbianus*) and which normally inhabit a 100-mile-wide band of woodlands and chaparral-covered coastal mountains extending inland from the Pacific Ocean. They are often mistakenly called "mule deer" but are a separate sub-species related to the California mule deer (*Odocoileus hemionus californicus*) found on the western flanks of the Sierra Nevada range and in the mountains of southern California, so called for their large ears. The most southerly population range of the black-tailed deer overlaps with that of the mule deer and hy-bridization between the two species is extensive.

Tule elk (*Cervus canadensis nannodes*), once abundant in the Bay Area numbering about half a million when the first Europeans arrived, are now only found in two herds, one on the Point Reyes peninsular and the other on Grizzly Island in the lower Sacramento River Delta, and are the largest native deer unique to California. In both cases, the herds having been all introduced about 40 years ago. They are also present in the Pacific coastal ranges from north of the bay to the Canadian border region.

Tule elk are a subspecies of the wapiti, which found over most of North America, from New Mexico in the south to British Columbia in the north, with smaller herds in Texas, North Dakota, and southern Canada. The name "wapiti" means "white rump" in the Shawnee and Cree languages, which is often all you would see of the animals in the wild (Canadian Encyclopedia). Another herd in

California is to be found in the Coyote Ridge Open Space Preserve near San Jose; still another resides in the Central Valley, at the San Luis National Wildlife Refuge and which was the site of the earliest herd to be bred in the 1870s–1890s when they were thought to have been extirpated, that is, locally extinct. They would have been the typical prey species for the grizzly bear, the black bear, and the cougar. Today, the Point Reyes herd is surrounded by fences and thus no longer preyed upon. Recently, however, a proposal from the National Park Service, which manages the 28,000 acres of farmland on Point Reyes, to cull some of this population in order to promote more tourism, has been sharply criticized by many conservation groups (Fimrite, 2019).

European wild boar (*Sus scrofa scrofa*) were introduced into Monterey, California, for hunting during the 1920s; its domesticated descendant, common swine or pigs were brought over by Spanish settlers during the 1700s and many have since become feral and the present-day population is a wild boar/swine hybrid (Mayer and Brisbin, 2008; Woodward and Quinn, 2011).

Mid-sized and small prey mammals are commonly seen in all parts of the Bay Area. Some examples are: the California ground squirrel (*Otospermophilus beecheyi*); the Western grey squirrel (*Sciurus griseus nigripes*); the fox squirrel (*Sciurus niger*), which appears to have been introduced to Southern California from the east coast around 1904 (King, 2004; Ortiz and Muchlinski, 2014), and which now is the predominant species in the Bay Area over the Western grey squirrel; native dusky-footed woodrats (*Neotoma fuscipes*); the brown rat (*Rattus norvegicus*, introduced 1750–1755; Norwak 1999); the more common black rat (*Rattus rattus*), observed often in yards and may inhabit attic space (so-called "roof rats"), and which was probably introduced during the early to mid-1800s and whose mtDNA is more closely related to rats from South and South-east Asia (Lantz, 1909; Conroy et al., 2012); the California vole (*Microtus californicus Rattus*, the San Francisco Bay Area subspecies); mice (*Mus musculus domesticus*); the ornate shrew (*Sorex ornatus*); moles, including the American shrew mole (*Neurotrichus gibbsii*), the broad-footed mole (*Scapanus latimanus*), having a subspecies, *S. latimanus parvus*, on the island of Alameda in the San Francisco Bay; and the coast or Pacific mole (*Scapanus orarius*); rabbits, including the black-tailed jackrabbit (*Lepus californicus californicus*) and the brush rabbit (*Sylvilagus bachmani*).

Another important rodent that claims the Bay Area as home is the Point Reyes mountain beaver, (*Aplodontia rufa phaea*), whose habitat is, unsurprisingly, Point Reyes. It is a subspecies of the Mountain beaver (*Aplodontia rufa*) and is also to be found further north along the coast, at Point Arena. As with many of the small- to mid-sized mammals, their territories have been subject to continuous encroachment by people, farming, and urbanization, and thus their population numbers and range have significantly decreased over the past 250 years or so.

One bane of Bay Area residents is the Botta's pocket gopher (*Thomomys bottae*); in their natural habitat, they create a network of tunnel systems that provide protection and a means of collecting food. However, when they reside in peoples' gardens, they can cause havoc and destroy lawns, small shrubs, and entire horticulture.

Marsupials

The only marsupial found in the Bay Area is the Virginia opossum (*Didelphis virginiana*). This ancient pouched mammal lives for only about 3 years but breeds when it is six (female) or eight (male) months old up to three times a year. They are generally nocturnal but are a familiar sight even in urban environments, especially when food is left outside for pets; they are omnivorous (Krause and Krause, 2006). The tail is prehensile and young opossums use this to grip onto the mother's back. As described in Chapter 5, the opossum's presence in North America resulted from the great American biotic interchange just under 3 million years ago.

Birds

Land Birds

For reasons of space, we cannot describe each bird species found in the Bay Area and we suggest the reader finds other more comprehensive resources, such as the many ornithological handbooks available to the general public. Here we describe some of the birds that are of particular significance to the ecology of the Bay Area.

The bird that has captured the most attention in the minds of Californians during the past few decades has been the re-introduction of the largest bird of prey in north America, and the largest of the north American New World vultures, the California condor (*Gymnogyps californianus*), second only in size to the Andean condor (*Vultur gryphus*).

Extinct in the wild since 1987, the then-captive population of some 22 individual birds were the subject of an intensive captive breeding program, the California Condor Recovery Plan, led by the San Diego Wild Animal Park and the Los Angeles Zoo (CDFW, nd-c) (Fig. 8.1). One innovative feature of the program was that no humans were allowed to be seen by any hatchlings, to not become habituated to humans; the condor chicks were fed small portions of meat by a puppet condor consisting of the head and neck, within which the scientist/keeper's hand and arm were hidden. Mixed-age California condors were released in 1991 and 1992 in California at Big Sur, Pinnacles National Park, and Bitter Creek National Wildlife Refuge (southwestern San Joaquin Valley), and the first breeding pairs were observed establishing a nest in 2006. This is due to the late age at which they begin to reproduce (6 years old) as well as the fact that condors will only raise a single chick every other year (FWS, 2007).

According to the latest information from the United States Fish and Wildlife Service, there are approximately 160 California condors present in Central and Southern California (FWS, 2020). However, recent wildfires in California may have contributed to loss of some of these magnificent birds, including at least one chick (Rubenstein, 2020).

FIGURE 8.1   California Condor. The birds were photographed at the San Diego Zoo, which was a key part of the effort to bring the condor back from the brink of extinction. Since the program began, the population of condors has grown from 22 to more than 500. Photograph reproduced with permission from Dr. Cynthia Wikler.

The turkey vulture (*Cathartes aura*), sometimes called a turkey buzzard, is another member of the New World vultures and is the most common scavenging bird seen in the skies of the San Francisco Bay Area; as would be expected, they are often seen soaring and circling up using updrafts high into the sky to better see over the vast plains and hills of northern California. They are also to be found in the middle of the highway, a small group tearing at the carcass of road-kill.

The Bay Area has a large number of native birds of prey, in particular since the environment provides habitat for so many typical prey. These include the great horned owl; the barn owl; burrowing owls, the bald eagle and golden eagle, red-tailed hawk, northern harrier or marsh hawk, and white-tailed kite (Robbins et al., 1983).

The crow family in the Bay Area is distinguished by the common raven (*Corvus corax*), the American crow (*Corvus brachyrhynchos*), and the California scrub jay (*Aphelocoma californica*), all of which are predominantly woodland and scrub birds (Robbins et al., 1983).

The San Francisco Bay Area is a critical stopover point for migratory birds along the Pacific Flyway migration route, in particular, the shallow waters of the bay and the Sacramento Estuary, which are a welcome rest point for shorebirds and waterfowl, as well as for sparrows and thrushes on the land, and which can number of one million birds.

Some of the other native birds and waterfowl are listed here: the American goldfinch (*Carduelis tristis*), the American robin (*Turdus migratorius*), Brewer's blackbird (*Euphagus cyanocephalus*), red-winged blackbird (*Agelaius phoeniceus*), California towhee (*Pipilo fuscus*), dark-eyed junco (*Junco hyemalis*), the northern mocking bird (*Mimus polyglottos*), great egret (*Casmerodius albus*), mallard (*Anas platyrhynchos*), Canada goose (*Branta canadensis*), and hummingbirds, including Anna's hummingbird (*Calypte anna*) (Robbins et al., 1983). The European or common starling (*Sturnus vulgaris*) is an invasive species, but giant flocks are to be seen in the early evenings, where they gather before nesting prior to sunset.

The Californian turkey, *Meleagris californica*, is an extinct species of turkey indigenous to the Pleistocene and early Holocene of California. It became extinct about 10,000 years ago (Bocheński and Campbell, 2006). The present Californian wild turkey population derives from wild birds re-introduced from other areas by game officials, a previous introduction program dating from between the 1920s and early 1950s having been unsuccessful. The Rio Grande wild turkey (*M. gallopavo intermedia*) was introduced throughout California during the 1960s and 1970s, whereas the Eastern wild turkey (*M. g. silvestris*) was released along the northern coast and has since formed a hybrid species with the Rio Grande subspecies (Cal. Dept. Fish Game, 2005).

Two birds frequently seen around the bay are the brown pelican (*Pelecanus occidentalis*) and the double-crested cormorant (*Phalacrocorax auritus*), both of which feed on fish by dive-bombing into the water from a height of usually about 5–10 m. They are often observed fishing together but since they are adapted to prefer different fish at different depths, there is little competition for food.

For further research, the reader is recommended to use some of the many well-produced hand-books available in the general press.

Seabirds and Shorebirds

One ecosystem in the Bay Area that we have not yet discussed is that of the perimarine environment, the shoreline and the areas adjacent to them. It is here that numerous gulls and other seabirds compete for food and nesting space. There are many recognized species within the gull group (Laridae) although genetic analysis often demonstrates hybridization between "species" and that most have up to 98.7% DNA in common (Paton et al., 2003; van Tuinen et al., 2004; Pons et al., 2005). The following lists the more common gulls seen in the Bay Area, not only at the seashore, but also farther inland, where they may be seen raiding trash heaps up to 200 km from the ocean (Akerman and Peterson, 2017).

Included in no particular order are the Western gull (*Larus occidentalis*), the California gull (*L. californicus*), the glaucous-winged gull (*L. hyperboreus*), the American herring gull (*L. smithsoniansus*), Theyer's gull (*L. glaucoides thayeri*), Heermann's gull (*L. heernmani*), and the black-headed gull (*Chroicocephalus brunnicephalus*). Some eat brine shrimp, grasshoppers, seal afterbirths, fish, squid, carrion, small mammals, and the eggs and chicks of other seabirds—even those of their own kind.

Other smaller seabirds include the black-legged kittywake (*Rissa tridactyla*) and the California least tern (*Sternula antillarum*), both of which spend more time at sea compared with that of the gulls. On the other hand, the waders or shorebirds spend all their feeding time scurrying along the flat beach just along the tidemark, darting here and there looking for worms, small crustaceans, and other arthropods. Examples of such waders are the western sandpiper (*Calidris mauri*), the American avocet (*Recurvirostra americana*), the black oystercatcher (*Haematopus bachmani*), and the western snowy plover (*Charadrius nivosus nivosus*), the latter which is threatened (Audubon California, nd).

Reptiles

Snakes

Let us begin with some of the snakes (Infraorder Serpentes) that inhabit the San Francisco Bay Area. The relatively warm and, until recently, previously wet climate, offered a multiplicity of welcome habitats for small birds, mammals, reptiles, and invertebrates, the common prey for all snakes.

Garter snakes are some of the more common small (<1 m) snakes found almost everywhere in the Bay Area, each species specialized to a particular habitat (California Herpes, 2020a). The California red-sided garter snake (*Thamnophis sirtalis infernalis*) is such an example. The prey of the San Francisco garter snake (*Thamnophis sirtalis tetrataenia*) is generally red-legged frogs and juvenile bullfrogs. However, San Francisco garter snakes are one of the few animals capable of ingesting the toxic California newt (*Taricha torosa*) without incurring sickness or death. Its relative, the Coast garter snake (*Thamnophis elegans terrestris*) consumes slugs.

Despite its name, the aquatic garter snake (*Thamnophis atratus*) lives throughout the San Francisco Bay Area, rivers, and hills alike; a subspecies *Thamnophis atratus zaxanthus*, the Diablo aquatic garter snake, is to be found in the hills of the East Bay as far south as the southern reaches of the Diablo range (San Benito Co.), and as a separate population from Monterey south to the Sierra Madre mountains (Santa Barbara Co.). Their diet includes fish, frogs, and toads. The range of the Santa Cruz aquatic garter snake (*T. atratus atratus*) is confined to the northern San Francisco Peninsular south to Santa Cruz and the Monterey Bay; its diet includes minute fish and amphibian larvae.

The Alameda whipsnake (*Masticophis lateralis euryxanthus*, also known as the striped racer) is frequently mistaken for the aquatic garter snake in that, in the Bay Area, its upper body is predominantly colored black and has two lateral yellow stripes along the length of the body. In contrast, the garter snake is similarly colored but has an additional yellow stripe running all the way down its back. Its diet includes live animals such as insects, lizards, snakes, birds, and even small mammals (Jennings, 1983; Swaim and McGinnis, 1992; Swaim, 1994; Stebbins 2003). It is endemic to California and is currently listed as endangered, since its habitat is being destroyed by industrial and urban development; it is now mainly found in the area of southern Alameda County, northern Santa Clara County, and western San Joaquin County, and in the southeastern Bay Area.

Another common snake often found sunning itself on the tarmac of a quiet byway, is the Western yellow-bellied racer (*Coluber constrictor mormon*). They are predominantly found in the foothill grasslands, brushlands, and moist environments throughout the Bay Area. They eat mice, fledgling birds, and lizards and are nonvenomous (Murray, 2004).

The California rattlesnake (*Crotalus oreganus*), also going by the name black diamond rattlesnake for its distinctive dorsal pattern, is not uncommon in the Bay Area; it is to be found predominantly in the hills and undeveloped valleys that surround the residential suburban torus that surrounds the San Francisco Bay.

One snake that is becoming increasingly rare in the Bay Area is the California kingsnake (*Lampropeltis getula californiae*). They are found under logs and rock outcrops throughout the region and their prey included other snakes, (including rattlesnakes) lizards, birds, eggs, and small rodents. They restrain and kill their prey as a constrictor, wrapping their body around the prey and suffocating it before ingesting the animal head-first. They are immune to rattlesnake venom.

Lizards

There are at least 20 species of native lizards (Order Squamata) that inhabit all the different ecosystems of the Bay Area; there are also a few introduced species (California Herps, 2020b). They include the alligator lizards (*Elgaria sp.*), a number of legless lizards (*Anniella sp.*), fence lizards (*Scaliporos occidentalis sp.*), and the California whiptail (*Aspidoscelis tigris munda*). Introduced species include geckos and anoles.

The most commonly seen, usually sunning themselves on a rock in backyards, are the California alligator lizards (*Elgaria multicarinata multicarinata*) which are a subspecies of the southern alligator lizard (*Elgaria multicarinata*). The term "alligator"; refers to the presence of supporting bony structures in their dorsal and ventral scales, as occurs in alligators. They can reach a size of over 30 cm including the tail (Stebbins, 2003).

Amphibians

Amphibians (class Amphibia) include salamanders, newts, frogs, and toads. The largest of the Bay Area amphibian is the California giant salamander (*Dicamptodon ensatus*) with a habitat ranging from Santa Cruz County to Mendocino County. The adult California giant salamander can reach from between 17 and 30.5 cm (6.7–12 inches) in total length (including tail) and is one of the only salamanders that vocalize (Hogan, 2008). Of particular interest, this is one of the many species of salamanders which can, under certain environmental conditions, retaining the larval form whist becoming sexually mature; this is termed "neoteny" and is considered in the tenets of biological sciences to be one way that evolution to adapt to changing conditions and speciation may occur, including in humans (Gould, 1977; Kucera, 1997). A salamander that may be familiar to the reader that is also neotenous is the Mexican axolotl (*Ambystoma mexicanum*) and which has been frequently kept in aquariums as a pet.

A salamander frequently spotted in the damp woodlands in the hills and mountains that surround the San Francisco Bay is a lungless salamander, the common ensatina, *Ensatina eschscholtzii*, and which has been the subject of continuous scientific study at the University of California at Berkeley since the 1940s. It was one of the first animals to be identified as a ring species, that is, the subspecies are distributed around a geographic feature and they eventually meet up at either end of the ring as two non-interbreeding species whereas those subspecies found in adjacent habitats "around the ring" could interbreed in what is termed a "hybrid zone" (Stebbins, 1949; Brown and Stebbins, 1964). Briefly, Stebbins determined from his studies of coloration and morphology that the seven different groups of salamander which encircled the California Central Valley were most likely each a different subspecies of the ensitina salamander and that the two species at the southern end of the range in the Tehachapi Mountains, *E. e. eschscholtzii* (western population) and *E. e. croceater* (eastern populations) could not interbreed. (Further south, in the Laguna Mountains, Santa Rosa Mountains, San Jacinto Mountains, and the San Bernardino Mountain ranges, another non-interbreeding population is found: *E. e. klauberi.*). These findings were later borne out by enzyme polymorphisms and genetic data (Moritz et al., 1992; Jackman and Wake, 1994; Wake, 1997).

The predominant subspecies of the Bay Area are *E. e. orogonensis*, which is found in San Francisco and counties north of San Francisco Bay, and *E. e. xanthoptica*, found in the San Francisco Peninsula, the South Bay, the East Bay, the Santa Cruz Mountains, and parts of Solano county (Stebbins, 1949). Interestingly, a small population of *E. e. xanthoptica* has been observed on the other side of the Central Valley on the western flanks of the Sierra Nevada; these most likely are descended from individuals who migrated from the Bay Area along the Calaveras or American Rivers during a period when the Central Valley was wetter (Stebbins, 1949, *op. cit.*; Wake, 1997, *op. cit.*). With the exception of the Lassen Gap zone, in almost all the regions where two subspecies' ranges adjoin each other, there is a zone of hybridization, which may only extend over a band of 730–2,000 m across (Alexandrino et al., 2005; Kuchta et al., 2009; Pereir and Wake 2009). This parameter conforms to the mean territorial range of the salamanders, about 19 m, in that the overlapping territories of neighboring salamanders would necessitate an intervening population numbering about 5–10 of each subspecies on either side of the hybridization zone (Stebbins, 1954).

Another salamander found in the Bay Area is the California tiger salamander (*Ambystoma californiense*); there are three distinct population segments in California, one in Santa Barbara County, one in San Joaquin County, the other in Sonoma County, just north of San Francisco (Shaffer et al., 2004). All are listed as federally threatened. Compared with the ensatina, it prefers open grassland habitat inhabiting vernal ponds and pools.

The California newt (*Taricha torosa*) inhabits much of the Bay Area, from the coastal counties north of San Francisco and south along the coastal ranges as far as the San Bernardino Mountains, east of San Diego. There is considerable evidence that the coloration pattern of the *Ensatina schscholtzii xanthoptica* is a mimic of the California newt (Kutcha et al., 2008).

The rough-skinned newt (*Taricha granulosa*) overlaps part of this range, as far south as Santa Cruz, but is mainly found in the Pacific northwest coast and mountain ranges of northern California, Oregon, Washington, British Columbia, and Alaska. Its skin exudes a neurotoxin, tetrodotoxin (TTX), which prevents the flow of sodium ions through voltage-gated sodium channels in the cell membrane of nerve cells and thereby blocking nerve impulses (Striedter et al., 2013). Interestingly, as mentioned earlier, one of its major predators, the common garter snake (*Thamnophis sirtalis*) exhibits resistance to the TTX (Williams and Brodie, 2003).

The California red-legged frog (*Rana draytonii*), now an endangered species, was once a very common sight in the 19th century and was made famous in Mark Twain's *The Celebrated Jumping Frog of Calaveras County*. Nevertheless, it may still be seen in riverine and wetland habitats, such as the in San Francisco's Golden Gate Park. In 2015, the California State Legislature decreed it to be the "state amphibian" and it is now classed as a protected species in certain designated parks and regions, including the counties of Contra Costa, Alameda, and San Mateo (Hammerson, 2008; CDFW, 2014). It is a common food source for the endangered San Francisco garter snake in San Mateo county and northern coastal Santa Cruz county (CDFW, 2020a).

The American Bullfrog (*Lithobates catesbeianus*, but also known as *Rana catesbeiana*), is to be found throughout the Bay Area, predominantly in the East Bay and South Bay. Although native to North America east of the Rockies, is an invasive species to the western United States and was first sighted during the late 1890s. It preys upon birds, bats, rodents, frogs (even its own young), snakes, turtles, and lizards and thus may have deleterious ecological effects (McKercher and Gregoire, 2020). It prefers to live in permanently inundated wetlands and so constructive habitat modification by man, that is, converting these wetlands to be ephemeral, may help to control the Bay Area's bullfrog population.

The Sierran chorus frog (*Pseudacris sierra*) is found throughout the Bay Area; its habitat, as expected, stretches from about Santa Maria in the south, up the coastal ranges to the south of Oregon, and eastwards across the Central Valley to the Sierra Nevada and Northwestern Nevada, the Cascades, and Rocky Mountains of Idaho and western Montana (USGS 2780, nd; Stebbins, 1959; 1972; 2003). It is sometimes mistaken for the Pacific treefrog (*Pseudacris regilla*, also known as the Pacific chorus frog), however, the latter's range is limited to the northern counties of California, Oregon, Washington, British Columbia, and southern Alaska (Frost, 2014).

Fish

One cannot write a book about the San Francisco Bay Area without including the Pacific salmon (*Oncorhynchus sp.*). Their presence in the past not only provided nutrition to the Native American populations throughout northern California but also provides a significant input to the local economy. The predominant Pacific salmon in the Bay Area is the Chinook salmon (*O. tshawytscha*) and is divided into two Evolutionary Significant Unit (ESU) populations: the Fall-run and the Spring-run (CDFW, nd-d), so named for the season that they enter the Sacramento Delta to spawn. Many of the rivers in the San Francisco Bay had been blocked by dams and other physical barriers to upstream migration, but much has been done recently to allow the migration to be successful (CDFW, nd-d). They are considered threatened under the ESA.

The other important Pacific salmon in the Bay Area is the Coho (*O. kisutch*) that populates the Central California Coast ESU from Santa Cruz in the south to Humboldt County in the north (CDFW, nd-e). The populations have been in decline since the mid-20th century in particular to over-fishing and commercial development of the streams and rivers. The adults enter the Bay Area streams in the fall and the fry emerge in March of the following year. After 1 year in fresh water they migrate to the ocean (CDFW, nd-e). They are considered endangered under the ESA.

INVERTEBRATES

The word "invertebrate" is no longer used by biologists as a valid classification term and the organisms we will be describing below in fact are in many ways even more dissimilar to one another than mammals are to fish. The more correct terminology for these "animals without a backbone" organisms is "protostomes," which describes a particular sequence during embryonic development and which we do not need to describe here. In contrast, the echinoderms and "vertebrates" or chordates are referred to as "deuterostomes" (Martín-Durán et al., 2016). Therefore, although echinoderms are not protostomes, we will nevertheless include them in this section.

By almost any measure, the invertebrates are a fascinating group. In sheer numbers, there are more of them than any other animal group. They also have the greatest diversity of any group: they make up 95% of all animal species. Yet they are easily overlooked. They include insects, spiders, crabs, snails, clams, squids, octopuses, earthworms, leeches, jellyfishes, sea anemones, and many more. Most of these are represented in the Bay Area.

Invertebrates, whether aquatic, marine, or terrestrial, are important and often under-studied components of any habitat (Maffe, 2000). Few reports focus on them. Yet they serve as sensitive indicators of the overall health of the environment. Many invertebrates are important for other organisms. They act as pollinators, herbivores, scavengers, predators, and prey. Without the pollination of food crops by honey bees, human diets would be quite different. Unfortunately, the San Francisco Bay estuary is one of those areas in which these important organisms have been under-studied.

Terrestrial Invertebrates

Insects

The San Francisco Bay Area is a major urban area with more than 7 million people. Amazingly, it is also a hot spot for insect diversity. The intersection of those sets of organisms also means that the Bay Area is a hot spot for threatened species: more than half of the arthropod species listed as endangered by the US government are in the Bay Area (Dobson et al., 1997).

The insect diversity in the Bay Area results from two factors. First, California overall is a biological hotspot due to its geology, soil, and climate. Its latitude ensures a Mediterranean climate, and the elevation varies from sea level to the Sierras. It is split by the San Andreas fault so that part is on the Pacific plate and another part is on the North American plate. Second, the combination of the hills and Bay yields a multitude of microenvironments with different climates and soils.

Unfortunately, much of the Bay Area has been developed. Buildings and streets cover more than 40% of the developed land, and the habitat for insects has been lost. For example, about 43% of the butterfly species have been lost. In addition to loss of habitat, those areas that can support insects have been fragmented. As is the case with many species of wild animals, the inability to move from one area to another is highly stressful. Also, the quality of those undeveloped areas does not allow them to support a sufficient population of some insects. Invasive species have also put pressure on native insects. These might be insects that compete with the native insects or even invasive plant species that crowd out native plants and reduce food sources for insects. Ice plant (*Carpobrotus edulis*) changes the nature of the soil that it grows in so that there are fewer insects and other invertebrates.

Preventing exotic insects from gaining a foothold is the best strategy for dealing with invaders. However, it is not always possible, and so, additional strategies are necessary to eradicate non-native insects (Liebhold and Kean, 2019). Many nascent infestations fail because the insects cannot establish themselves in the new environment. They might not find suitable food or habitat or there might be too few insects to succeed. However, once the non-native insects get a start, controlling them in a forest is very challenging. The good news is that not every insect must be killed. The population only needs to be reduced below a threshold.

Some eradication programs have succeeded and other have failed. One example of a failed effort is the light brown apple moth (*Epiphyas postvittana*) in the Bay Area. Pesticides were not used because the infestation occurred in a populated area. Mating disruption was attempted, but the public came to believe that toxic substances were being used and they turned against the program. It was discontinued.

The SF Bay Wildlife maintains an excellent website that lists and describes many of the insects of the Bay Area (https://sfbaywildlife.info/species/insects. htm). Some insects are more welcome than others by the general public: bees, butterflies, and dragonflies. The Bay Area has about 90 species of native bees from five families (i.e., Apidae, Andrenidae, Colletidae, Halictidae, and Megachillidae). Honey bees (*Apis mellifera*) are an introduced species, originally from Europe. Other common bees include the Western bumble bee (*Bombus occidentalis*), yellow-faced bumble bee (*B. vosnesenskii* or *B. caliginosus*), California carpenter bee (*Xylocopa californica*), and green metallic bee (*Agapostemon sp.*). The European wool carder bee (*Anthidium manicatum*) is an introduced species that is spreading rapidly.

Dragonflies are a more primitive insect. They have four wings rather than two wings and halteres of more advanced flying insects (e.g., diptera). Some of the common dragonflies include common green darner (*Anax junius*), blue-eyed darner (*Rhionaeschna multicolor*), California darner (*Rhionaeschna californica*), variegated meadowhawk (*Sympetrum corruptum*), and cardinal meadowhawk (*Sympetrum illotum*).

About 144 species of butterflies are found in the Bay Area. The famous Monarchs (*Danaus plexippus*) now overwinter in the Bay Area and around Monterey in eucalyptus trees, a non-native species. Other species include the Western tiger swallowtail (*Papilio rutulus*), Western pygmy blue (*Brephidium exilis*), and painted lady (*Vanessa cardui*).

The Bay Area used to be home to over 100 species of ants. However, the introduced Argentine ant (*Linepithema humile*) arrived in the Bay Area in 1908 and has driven out most of the native ants in urban areas. The Argentine ants are very small and often invade houses in search of food. In citrus groves and vineyards, the ants tend aphids that produce honeydew from the plants. They protect the aphids from other predators. They are detrimental to plants that depend on ants to disperse their seeds (Christian, 2001). Some have suggested that a supercolony of Argentine ants stretches from Northern California to the Mexican border. However, genomic studies (of the ants) show that they are different and probably the result of the introduction of multiple clones of the ants (Ingram and Gordon, 2003). Nevertheless, the ants do form massive colonies that include billions of ants and multiple queens (Moffett, 2012). Other ant species include the bicolored carpenter ant (*Camponotus vicinus*), black carpenter ant (*Camponotus quercicola*), dark log ant (*Hypoponera opacior*), and red wood ant (*Formica integroides*).

Wasps are narrow-waisted insects that can sting. There are tens of thousands of species. While some live in colonies, most are solitary. They are valuable pollinators. Several species occur in the Bay Area, including the European paper wasp (*Polistes dominula*), California yellowjacket (*Vespula sulphurea*), blue mud wasp (*Chalybion californicum*), and Tarantula hawk (*Pepsis pallidolimbata*).

Beetles are in the order Coleoptera. They are different from most other insects in that their front wings have hardened into wing cases called elytra. With 400,000 species, beetles are an amazing successful and diverse group of organisms. They account for about 40% of all insects and 25% of all known animals. They are well-represented in the Bay Area with many species, including the seven-spotted lady beetle (*Coccinella septempunctata*), spotted cucumber beetle (*Diabrotica undecimpunctata*), and cobalt milkweed beetle (*Chrysochus cobaltinus*).

Grasshoppers have been around for 250 million years. These ground-dwelling insects have powerful rear legs that allow them to escape predators. Those in the Bay Area include the California sulphur-winged grasshopper (*Arphia behrensi*), the California rose-winged grasshopper (*Dissosteira pictipennis*), and the Jerusalem cricket (*Stenopelmatus fuscus*).

These are only a few of the insects in the Bay Area. There are also praying mantises, walking sticks, and cicadas. However, in recent years, scientists have become worried about the decline in the numbers of insects. The cause of the decline seems to be human activities (Simmons, 2019). A series of papers has sounded the alarm. The loss of insects will affect our food supply. Hallmann et al. (2017) used traps across Germany to determine the number and types of insects. They found that flying insects have decreased by 75% over the 27 years of study. Lister and Garcia (2018) documented a similar decline in arthropods in the soil. Mathiasson and Rehan (2019) showed a loss of a significant number of bee species. Beekeepers noted that they lost 40% of their hives in the winter of 2018–2019 (Neilson, 2019). The causes are thought to include decreasing crop diversity, poor beekeeping practices, loss of habitat, and indiscriminate use of pesticides. Infestation by a mite (*Varroa destructor*) has also harmed hives.

More groups than just bees are being lost. Simmons (2019) pointed out that insects are also being lost in the Bay Area. The monarch butterfly is a local favorite and a good example. Tourists flock to its overwintering grounds near Monterey. However, its numbers have been reduced by a shocking percentage. Other insect populations are also in deep decline.

Arachnids

Arachnids are a large class of invertebrates that feature spiders, scorpions, ticks, mites, harvestmen, and more. Most have eight legs, but some have other appendages that might look like legs. In addition, there are some mites that have fewer legs. Unlike insects, they also have no antennae or wings, and they have two main body segments rather than the three of insects.

Spiders

Spiders are very common essentially everywhere, and the Bay Area has its share. The western black widow (*Latrodectus hesperus*) is one of the few spiders in the US that are venomous to humans. Fortunately, they tend to be shy and retiring.

Contrary to several reports, brown recluse spiders have not been found in the Bay Area (Vetter, 2000). Tarantulas (*Aphonopelma iodius*) and false tarantulas (*Calisoga longitarsus*) are large, hairy, and not harmful. The cross orbweaver (*Araneus diadematus*) is a non-native species that builds large beautiful webs. The marbled cellar spider of daddy longlegs (*Holocnemus pluchei*) has a brown stripe on its ventral side. Another daddy longlegs is the longbodied cellar spider (*Pholcus phalangioides*); it is gray colored with an elongated abdomen.

Scorpions

The Bay Area has a number of scorpions (Buhler, 2018). They are rarely seen as they are nocturnal. They are also not dangerous unless the person stung happens to be allergic to the venom. Amazingly, they can most easily be found by using a black or ultraviolet light at night. The scorpions fluoresce brightly. They give birth to live young that ride on the mother's back early on. The most common species in the Bay Area is the California forest scorpion (*Uroctonus mordax*). Others include the California common scorpion (*Paruroctonus silvestrii*), sawfinger scorpion (*Serradigitus gertschi*), and the California swollenstinger scorpion (*Anuroctonus pococki*).

Worms

We are most familiar with earthworms, but there are many, many others. Even earthworms are not well studied in the Bay Area, and many of those here are imports. The native earthworms can still be found in undisturbed areas. Central California and the Coastal Ranges have 22 species in 13 genera of five families (Acanthodrilidae, Lumbricidae, Megascolecidae, Ocnerodrilidae, and Sparganophilidae). Only six of the species are native to the area. The rest are invaders. Reynolds (2016) has an extensive list of all of the species and their locations.

Many worms are parasites, and some parasitize humans. Nematodes, which are roundworms, occur in many species of fish. They appear on the intestines, liver, in the body cavity or in the flesh. They also live in seals, porpoises, whales, and dolphins. Their eggs escape in the feces of the mammal and are eaten by small crustaceans that are then eaten by fish or squid. The nematodes can also infect humans if they eat raw or unprocessed fish. Flukes and tapeworms are also common in the Bay Area fish and other animals.

Woodlice

These small crustaceans have segmented, dorso-ventrally flattened bodies with seven pairs of jointed legs. Females carry the fertilized eggs in their marsupium on the ventral side of her abdomen. The young emerge and seem to be a live birth, but they are really from eggs. There are thought to be 5–7,000 species worldwide. Most are useful in that they turn the soil like earthworms. However, they do eat some crops (e.g., strawberries) and are considered a pest if they invade homes. One common group is the pill bugs or rolly-pollies (*Armadillidium vulgare)* that are familiar to most children. They are common under rocks, leaves, or boards in the back yard. They were introduced to the Americas from Europe, but have thrived here. They breathe through gill-like slits and so require moisture to survive.

Millipedes and Centipedes

At first glance, millipedes and centipedes look like bugs with lots of legs. However, they are quite different. Millipedes have two pairs of legs per body segment. While a small number are predators, most eat decaying matter. Centipedes have only one pair per segment. They are essentially carnivorous.

There are a number of species of each in the Bay Area. The yellow-spotted millipede (*Harpaphe haydeniana*) is 5-8 cm when mature. It is black to olive green with patches of yellow on the sides. It has about 20 segments with 30 (males) or 31 (females) pairs of legs. In males, the 7th segment is modified for gonopods. They are valuable because they breakdown plant matter (especially redwood litter) into humus. *Xystocheir dissecta* is a species of flat-backed millipede. The first five to six segments are smooth, but those following have papillae. They can be light green or olive with some orange. They live in loose dirt, leaf litter, under rocks and in decaying logs. They are hard to see in the day, but fluoresce under UV light at night.

There are four groups of centipedes, but only three live in the Bay Area. Stone centipedes (*Lithobiomorpha spp.*) are small (usually less than 1 inch) and have 15 pairs of legs as adults. They are common in gardens and eat small insects. Soil centipedes (*Geophilomorpha spp.*) small and can have more than 60 pairs of legs. They live in the ground and eat subterranean insects. House centipedes (*Scutigeromorpha coleoptrata*) include only one species. They 15 pairs of long legs and are about 5 cm long. These are the only local centipedes that can bite, but their fangs are weak.

Snails and Slugs

Snails and slugs are both mollusks. Slugs have lost the shell that covers snails. All produce mucus to facilitate their movement and protect against desiccation. For a more extensive list, the reader is referred to the *Checklist of the Land Snails and Slugs of California* (Roth and Sadeghian, 2006).

Banana slugs have the name because of their general shape and bright yellow color. However, many of them are also green, brown, tan, and white. They can also grow to nearly 25 cm. They have a radula, which they use for feeding. By eating leaves, animal droppings, moss, and dead plant material, they produce soil humus and provide a valuable service. Two common species in the Bay Area are the Pacific banana slug (*Ariolimax columbianus*) and the slender banana slug (*Ariolimax dolichophallus*).

The most common snail in the Bay Area is the non-native European garden snail (*Helix aspersa*), but there are several native species (Ellis, 2008). The Monadenia land snail (*Monadenia infumata*) has a very dark shell and beautiful purple flesh. Moro shoulderband snails (*Helminthoglypta spp.*) are the most common native land snails in the Bay Area and are easily confused with garden snails. They are found mostly in undisturbed native ecosystems. The California lancetooth (*Haplotrema minimum*) eats plant material and other snails and slugs.

Marine Invertebrates

Many invertebrates are found in the Bay and Pacific Ocean (Swensrud and Platt, 2007; NPS, 2018). The numbers of different types of organisms are almost beyond counting. They can be best seen while snorkeling, scuba diving, or just looking at tide pools. The tide pools at the Fitzgerald Marine Reserve at Moss Beach, California, feature an extraordinary number of algae, crabs, mollusks, seastars, and others organisms. They are easily visible at low tide.

Here we will name a sampling of common species. For a more detailed account of marine invertebrates, we recommend this outstanding book *Intertidal Invertebrates of California* by Morris et al. (1980).

Sponges

The bodies of sponges have pores that allow water to circulate. They are filter-feeders and feed on single-celled organisms, detritus, and other material that they filter out of the water. The red beard sponge (*Clathria prolifera*) is a red or orange-brown colored sponge with many projections. Sponges are filter-feeders: they eat bacteria and detritus that they draw into their body. This sponge was originally from the Atlantic Ocean and arrived in the Bay in the 1940s. The yellow sponge (*Halichondria bowerbanki*) appears as a flat mass on objects. It eats plankton. It was also from the Atlantic, but arrived in the Bay in the 1950s.

Jellies

Jellies are graceful, beautiful, and sometimes dangerous. They go with the flow, literally. The current moves them along, but a few can also propel themselves to a degree. They vary in size from almost microscopic to specimens with tentacles of over 35 m. Among the many species found in the Bay and Pacific Ocean near the Bay are these. The moon jelly (*Aurelia sp.*) resembles a large dinner plate. It is translucent to whitish with short tentacles at the rim and four longer tentacles at the center. It eats plankton that is captured in the mucus on the tentacles and passed up to the mouth in the bell. The bell medusa (*Polyorchis sp.*) is globular and up to 4 cm tall. Tentacles hang from the rim. It eats zooplankton, which are transported to the mouth as in the moon jelly. This jelly can swim somewhat. Comb jellies (*Pleurobrachia sp.*) are transparent with eight rows of "combs" or cilia. It has two tentacles that collect food. These jellies bioluminesce in the dark. Their cilia aid in movement. The orange anemone (*Diadumene* sp.) is orange, flesh, or salmon pink in color and is also an introduced group.

Sea Anemones

Anemones look like underwater flowers (Gong, 2019). They have two lives: first, they form stationary polyps, and second, they produce eggs that become mobile planula larva. Their phylum Cnidaria also includes corals, sea pens, and gorgonians. All anemones are carnivorous. The common California genus

Anthopleura eats anything that gets caught in its stinging tentacles. The orange-striped green anemone (*Diadumene lineata*) attaches to various underwater objects. It is shiny green, olive, or olive brown with orange stripes. This non-native anemone was introduced years ago to the Bay, and it is now worldwide.

Flatworms

Flatworms belong to the phylum Platyhelminthes (Watkins, nd). Two of the three classes Cestoda (tapeworms) and Trematoda (flukes) are parasitic, and one, Turbellaria, is free-living. They are very simple and lack a respiratory or circulatory system. They absorb oxygen through their skin. Their mouth is at mid-body. Food enters the mouth and waste is expelled from the mouth also. The free-living flatworm *Notoplana acticola* is 2.5-7.6 cm long with two eyespots on the dorsal surface. It is brown or gray and eats small crustaceans, zooplankton, other worms, and dead animals. Flatworms are very colorful and graceful in the water.

Roundworms and Worms

Roundworms belong to the phylum Annelida, and they are both terrestrial and marine. *Megasyllis nipponica* is originally from Japan (Tighe, 2019). They live mostly in the mud, rocks, and other places at the bottom of the Bay. In the summer, they change radically. Adult worms move eggs and sperm to the rear of the worm and bud off a new worm. This schizogamy yields bright orange epitokes that swim around looking for partners. Upon finding a partner, they burst open to allow the sperm to fertilize the eggs. The native scale worm (*Halosydna brevisetosa*) is about 5 cm long with 18 pairs of dorsal scales. It can be red, tan, or brown in color. They are scavengers. Some live in the tubes of tube worms in a form of commensalism.

The leech *Branchellion lobata* is about an inch or so long with a sucker at each end. It feeds by attaching to sharks, fish, and squids and drinking their blood. They begin as males and later become female.

Like many invertebrates, the worms can also be quite strange. The echiuran *Urechis caupo* (the fat innkeeper worm) inhabits a U-shaped burrow in both intertidal and subtidal mudflats. It "plugs" the burrow and uses its body to pump seawater through the burrow so that it can capture food (Arp et al., 1992). Other invertebrates live with the worm and feast on its leavings (Parr, 2019). These include the clam (*Cryptomya californica*), the scale-worm (*Hesperonoe adventor*), the pea crab (*Scleroplax granulata* or *Pinnixa franciscana*), and the hooded shrimp (*Betaeus longidactylus*). Strong storms can break up the sand in which the worms live, and occasionally, thousands of them can be seen stranded on beaches in Northern California.

Invasive species can travel in a surprising way. For example, bait worms (*Nereis virens*) are harvested in brown seaweed (*Ascophyllum nodosum*) from Maine and shipped many places, including the Bay Area. Once the worms are used, the boxes are often tossed into the water. Unfortunately, the boxes still contain other organisms. Haska et al. (2012) found 13 species of macroalgae and 23 species of invertebrates in the boxes. Two potentially toxic microalgae

(*Alexandrium fundyense* Balech and *Pseudonitzschia* multiseries (Hasle)) were also found. *A. fundyense* was found in the Bay Area.

Nudibranchs

The words "sea slugs" might conjure up a sea-going version of the slugs we see on land. Nothing could be further from the truth. Sea slugs come in many colors, and many are festooned with cerata (soft projections) (Ueda and Agarwal, nd). As a result, common nudibranchs, sea hares, and sapsucking slugs are some of the most beautiful of sea creatures. Over 3,000 species of nudibranchs are known (Bergamin, 2014). There are two forms of nudibranchs. Dorids are large, round, and flat. Aeolids are smaller with lush, feathery gills called cerata. The native nudibranch *Hermissenda crassicornis* is white with a blue line on each side. Its cerata are orange-brown with white tips. They eat hydroids that have nematocysts. The nudibranchs are immune to the stinging cells and, in fact, store them to reuse in their own defense. *Phidiana hiltoni*, the killer sea slug, has spread north from Monterey as the climate in the Bay Area has warmed. Sea hare (*Phyllaplysia taylori*) is a primitive marine gastropod lives primarily in eel grass (*Zostera marina*) and feeds upon the sponge larvae and diatoms that settle on the eel grass blades.

Clams

Clams are bivalves. Their two calcerous shells are connected by two adductor muscles, and they have a power foot for burrowing into the sand. Unlike oysters and mussels, they do not attach to a substrate. The term "clam" is typically used to describe those bivalves that are edible. They are all filter feeders.

The Bay has many types of clams. The overbite clam (*Corbula amurensis*) is white, tan, or yellow. The name comes from the fact that one shell is somewhat larger than the other. Originally from the Far East, it is now well established in the Bay and Delta since 1986. The Atlantic softshell clam (*Mya arenaria*) is white or gray with a dark siphon. As the name indicates, this clam is non-native. It was introduced before 1874. Interestingly, it can use anaerobic respiration, which is rare among mollusks. The gem clam (*Gemma gemma*) is white or buff. It is also a non-native species. It arrived in the late 19th century. Young ones are brooded by the mother in the mantle folds. Clearly, the Bay has received many introduced species. One of the best documented is the Asian clam *Potamocorbula amurensis*, which was introduced into the Bay in 1986, and within 2 years, it had firmly established itself throughout the Bay (Carlton et al., 1990). Now some areas contain 10,000 clams per cubic meter. It is primarily subtidal and can thrive in any level of salinity in the Bay.

Invertebrates, including clams, can also be very destructive. For centuries, the shipworm *Teredo navalis* was the bane of wooden ships and other wooden structures in the water. Although it is called a "worm," it is actually a clam. In 1920, a wharf belonging to the Union Oil company collapsed near the Carquinez Strait. The clams had eaten much of the wooden material and left it as a weakened sponge-like structure. Eventually, it was not strong enough to support itself and it collapsed.

Mussels

The bay mussels *Mytilus trossulus* and *M. galloprovincialis* are bluish black. These filter-feeders eat detritus and microscopic plants and animals. *M. trossulus* is a native species, and *M. galloprovincialis* was introduced from the Mediterranean. They form dense clusters on rocks, piers, and other substrates that are in calm waters. The green mussel *Musculista senhousia* is dark with brown or purple and green bands. This nonnative forms thick mats in the mud.

Oysters

Oysters are filter-feeders that eat plankton, bacteria, and detritus. They are found throughout the intertidal zone. *Crassostrea conchaphila* is a common native oyster in the Bay. The shell is 5-8 cm long with a wavy edge. Oysters have been important in the Bay for hundreds of years. The Native Americans used them as a handy source of food and built mounds with the left-over shells. Unfortunately, native oysters are in a serious decline in the Bay. In the 1850s, the Olympia oyster (*Ostrea lurida*) was once very common all along the west coast (Green, 2014). However, the oyster never really did well in the Bay, and the silt from the gold mining killed off the oyster industry. Since that time, oyster farming in the Bay has never done that well.

Sea Stars

Sea stars (also known as starfish) are echinoderms. About 1,500 species are known, and they are found around the world. They feature a central disc and usually five arms, but some species have many arms. Their colors vary. Although they seem peaceful, they are actually predators. The pink bay star (*Pisaster brevispinus*) is pink. It eats mussels and other bivalves and sand dollars. It is found near the Golden Gate since it needs a high level of salinity. The brittle star has five thin very flexible arms. It is brown or gray and eats detritus and plankton.

Since 2013, sea stars have been suffering from a disease called sea star wasting disease (Jaffe et al., 2019). More than 20 species have been affected. The disease causes necrotic lesions, twisted rays, ray loss, and death. It seems to be viral, but the causative agent has not been identified.

Jaffe et al. (2019) examined the disease in *Leptasterias* spp. This small sea star broods its young rather than releasing them as planktonic larvae, and thus, they might be more susceptible to a viral disease. In 2010, this sea star was common in the Bay. In 2016, few could be found. The researchers found that this sea star suffered symptoms similar to larger sea stars. The loss of this species is not a good sign for the general health of the Bay. Twenty other species are also in decline.

Harvell et al. (2019) studied the loss of the common predatory sunflower sea star (*Pycnopodia helianthoides*) throughout its traditional range along the West

Coast. Sea star wasting disease is responsible, and they found that its outbreaks were associated with unusually warm surface temperatures.

Crustaceans

Crustaceans are a very large group that includes crabs, lobsters, crayfish, shrimp, krill, woodlice, and barnacles. About 67,000 species are known. All have exoskeletons that must be shed for the animal to grow.

Crabs

Crabs are decapod crustaceans with a thick exoskeleton and a pair of pincers. Crabs live in both salt and fresh water around the world. Multiple species live in the Bay Area. The Dungeness crab (*Cancer magister*) is the best-known crab in the Bay Area. It is a favorite dish for residents and tourists alike. The Bay is a nursery for the Dungeness, and crabbing in the Bay is illegal. The native spider crab (*Pyromaia tuberculate*) eats algae and other plants. Sponges, algae, and other organisms grow on the crab's body and legs and give it camouflage.

Shrimp

Shrimp are decapod crustaceans with elongate abdomens. In this way, they are more similar to lobsters than crabs. Their many swimmerets allow them to swim well. There are thousands of species, and they are an important food crop.

The Bay contains multiple species. The Korean shrimp (*Palaemon macrodactylus*) is reddish with a prominent rostrum. These shrimp are omnivores and often scavenge. This non-native species arrived in the Bay in 1957. Bay shrimps (also called grass shrimp) (*Crangon* spp.) are semi-transparent with black spots. They eat smaller shrimp, amphipods, clams, and plants. This is a native species. They are caught now mostly for bait. While Native Americans likely enjoy the bay shrimp, early Spanish and Americans ignored them until Chinese workers began using them. Bay shrimp are sensitive to salinity and temperature. They like brackish water with a salt concentration of 14–24 parts per thousand and about 65 °F. They will migrate for miles to find this combination. They are protandrous hermaphrodites. That is, they begin life as males for a year and then change to females. The number of shrimp in the Bay varies considerably from year to year. For example, the numbers of *C. franciscorum* in 1996 was 20 times that of 1980. Snapping shrimp (family Alpheidae) create an audible sound that can be heard in the Bay. In fact, it is second only to the echolocation clicks of sperm whales in volume. They use specialized claws to make the sounds, stun prey, and communicate. The Bay ghost shrimp (*Neotrypaea californiensis*) is pale in color and grows to about 13 cm. It is a filter feeder that lives in a burrow and reworks the Bay bottom somewhat like earthworms do for soil. They disturb oyster beds.

Octopi and Squid

Squid and octopi are cephalopods (phyllum Mollusca). They can change colors quickly. While octopi were common in the Bay in the 19th century, today they are rare. Two species of octopi live in the Bay. Those with an arm span of over 60-90 cm are a giant Pacific octopus (*Enteroctopus dofleini*). They can grow to nearly 0.9 m. Those octopi with an arm span less than 60-90 cm are either the East Pacific octopus (*Octopus rubesens*) or a young *E. dofleini*. The Humboldt squid (*Dosidicus gigas*) periodically arrives by the thousands in Northern California. They are about 7-27 kg. Most squid caught for eating are California market squid or opalescent inshore squid (*Doryteuthis opalescens*). They are found near shore and reach a length of about 25 cm.

PLANTS

The San Francisco Bay Area is classified with regard to plants as temperate warm-summer Mediterranean Climate Zone (NOAA, 2020), where most the Bay Area falls into climate zones 14 through 17, with zone 7 south of San Jose (Sunset, 1995). Native plants, such as the Montara Manzanita (*Arctostaphylos montaraensis*) (Fig. 8.2) or the Antioch Dunes evening primrose (*Oenthera deltoides* subsp. *howellii*) are listed as critically endangered species, mainly due to land-use change, new developments and off trail/road walking and vehicle (e.g., motorcycles, mountain bikes) habitat degradation (California Native Plant Soc., 2017). The predominant trees are California live oak (*Quercus agrifolia*) (Fig. 8.2), California scrub oak (*Quercus berberidifolia*), which make up most of the Chaparral-type woodlands, and the coast redwood (*Sequoia sempervirens*), which naturally occur in damper environments, particularly in valleys. The acorns from these oaks were a large source of carbohydrate and protein for at least twelve Native American tribes in the historic Bay Area. The coast redwood's habitat is limited to the wet and foggy Coastal mountain ranges of northern California and Oregon and is distinct from its cousin, the giant sequoia (*Sequoiadendros giganteum*), of the western Sierra Nevada range, both may be found as ornamental trees in many Bay Area gardens (Fig. 8.2). Another native tree is the California buckeye (*Aesculus californica*), also called the California horse chestnut, for its chestnut-like conkers, and which is found in the cooler coastal and foothill environments of Californian; it is drought and salt tolerant (SelecTree, 2020).

One significant botanical resident that has become almost endemic to the Bay area is the many different types of eucalyptus species (*Eucalyptus spp.*, in particular the blue gum, *E. globulus*, and *E. camaldulensis*) that dot the hillsides in large groves; these are an invasive species that had been introduced from Australia to San Francisco in the 1850s with the expectation that they would provide a rapidly growing source of timber and fiber for paper as well as for ornamental uses (Santos, 1997; Groenendaal, 1983). However, the wood is so dense that harvesting and lumbering is extremely energy-intensive and thus they were left to their own devices. Efforts are underway to clear at least some of the most prolific and old stands (Wolf, 2015).

The ornamental species planted by landscapers, businesses, and homeowners are too numerous to mention here, given that our aim was to describe the native

FIGURE 8.2   Common Plants in the Bay Area (clockwise from the upper left) Manzanita (*Arctostaphylos* sp.), redwoods (*Sequoia sempervirens*), coyote bush (*Baccharis pilularis*), wetlands, poison oak (*Toxicodendron diversilobum*), and acorns of the coast oak (*Quercus agrifolia*).

environment. The reader should refer to other works that give more weightage to such a topic.

The historic shorelines of the Bay were characterized predominantly by marshland, wetland, and tidal flats, both originally were inundated on a daily basis by tidal action (Fig. 8.3). However, since the 1950s, much of the low-lying shoreline has been developed, both for housing and for industrial use. This has resulted in a transition from the once-native pickleweed (*Salicornia europaea* and *S. virginica*) to the invasive cordgrass (*Spartina*), the more salt-tolerant native salt grass (*Distichlis*), and the invasive Mediterranean saltwort (*Salsola soda*) and Australian saltbush (*Atriplex semibaccata*) (Baye, 2006; Goman et al., 2008).

Eelgrass (*Zostera marina*) along the shore of Point Molate Beach Park (Richmond, California); here it occupies over 20 hectares of the shallow seabed and provides a home to Taylor's sea hare (*Phyllaplysia taylori*), a type of sea slug gastropod mollusk. Eelgrass also provides a spawning habitat for the Pacific herring (*Clupea haengus pallasi*) and outmigrating juvenile salmon (*Oncorhyncus spp.*) (Ort et al., 2012). Below the surface, at an average depth of about 4 meters, the soil is rich in carbon, providing a large storage system of carbon in the Bay Area (Poppe and Rybcyk, 2018).

One significant large organism that also occupies the shallow coastal waters is the giant kelp (*Macrocystis spp.*), a marine macroalga whose photosynthesis provides nourishment to a large ecosystem. The kelp forest comprises stripes (stalks) and fronds (leaves) up to 200 m in height and which provide shelter to hundreds of species (Foster and Schiel, 1985). Under optimal conditions, they may grow at a rate of 30–60 cm per day (National Ocean Service, 2013; 2020).

INVASIVE SPECIES

Invasive species can have a detrimental effect on native species. The movement of humans from continent to continent provides an excellent opportunity for plants and animals to move from one region to another. For example, the Chinese mitten crab arrived in the San Francisco Bay in about 1992. Since then, this native of Central Asia has spread throughout the Bay. By 1999, parts of the tidal flats in the South Bay had 30 burrows/square meter. They inhabit both salt and fresh water environments. While the numbers of mitten crabs have cycled up and down, they are very difficult to control. There is concern that they will compete with native crabs. In addition, since they are a burrowing species, there is concern that they will dig holes into levees and damage them.

Plants can also be invaders. For example, several species of invasive cordgrasses (*Spartina* sp.) are found in the San Francisco Bay, and they may compete with native species *S. foliosa*. *S. alterniflora*, and *S. foliosa* easily form hybrids, and they are found mainly in the South and Central Bay (Ayres et al., 2004). They spread by rhizomes, and the hollow stems of smooth cord grass (*S. alterniflora*) grow from 0.6-1.3 m tall. It was introduced to the West Coast in an

FIGURE 8.3 Wetlands. Wetlands once surrounded the Bay. However, much of the wetlands have been lost to development over the last 150 years. More recently, efforts have been underway to restore wetlands.

attempt to prevent erosion, but it is now considered an invasive species. *S. anglica, S. densiflora,* and *S. patens* have had a more limited spread in the Bay. The exotic species outcompete the native species. They also change the mudflats to meadows and destroy the habitat of endangered species, such as the salt marsh harvest mouse. Moreover, the exotic species are detrimental to migrating shore birds.

REFERENCES

Abadía-Cardoso A., Freimer N.B., Deiner K., Garza J.C. (2017) Molecular population genetics of the Northern Elephant Seal Mirounga angustirostris. *Journal of Heredity* 108(6): 618–627.

Akerman J.T., Peterson S.H. (2017) California gull diet, movements, and use of landfills in San Francisco Bay, *Tideline* 40: 1–2.

Alexandrino J., Baird S.J.E., Lawson L., Macey J.R., Moritz C., Wake D.B. (2005) Strong selection against hybrids at a hybrid one in the *Ensatina* ring species complex and its evolutionary implications. *Evolution* 59: 1334–1347.

Arp A.J., Hansen B.M., Julian D. (1992) Burrow environment and coelomic fluid characteristics of the echiuran worm *Urechis caupo* from populations at three sites in northern California. *Marine Biology* 113: 613–623.

Audubon California (nd) California snowy plover. https://ca.audubon.org/westernsnowyplover, accessed September 28, 2020.

Ayres D.R., Smith D.L., Zaremba K., Klohr S., Strong D.R. (2004) Spread of exotic cordgrasses and hybrids (Spartina sp.) in the tidal marshes of San Francisco Bay, California, USA. *Biological Invasions* 6: 221–231.

Barrat J. (2013) Suburban raccoons more social yet dominance behavior remains that of a solitary animal. *Smithsonian Insider.* https://insider.si.edu/2013/07/suburban-life-does-not-alter-solitary-ways-of-the-raccoon/, accessed September 27, 2020.

Bay Area Puma Project (nd) *Felidae Conservation Fund.* Archived from the original on March 23, 2010, accessed September 25, 2020.

Baye P.R. (2006) Selected Tidal Marsh Plant Species of the San Francisco Estuary: A Field Identification Guide, San Francisco Estuary Invasive Spartina Project (California Coastal Conservancy) www.spartina.org.

Bergamin A. (2014) Nudibranchs, kings of the tidepool, command an audience. *Bay Nature.* https://baynature.org/article/nudibranchs-kings-tidepool/, accessed September 30, 2020.

Beschta R.L., Ripple R.J. (2009) Large predators and trophic cascades in terrestrial ecosystems of the western United States, *Biological Conservation* 142: 2401–2414.

BirdLife International (2007) Species factsheet: California Condor *Gymnogyps californianus.*

Bocheński Z.M., Campbell K.E. Jr (2006) The Extinct California Turkey, *Meleagris californica,* from Rancho La Brea: Comparative Osteology and Systematics. *Natural History Museum of Los Angeles County* 330: 1–92.

Brown C.W., Stebbins R.C. (1964) Evidence for hybridization between the blotched and unblotched subspecies of the salamander. *Ensatina eschscholtzii. Evolution* 18: 706–707.

Buhler B. (2018) There are so many scorpions. *Bay Nature,* Winter 2018. https://baynature.org/article/there-are-so-many-scorpions/.

California Department of Fish and Game (2005) *Guide to Hunting Wild Turkeys in California.* Sacramento, California Department of Fish and Game, Wildlife Programs Branch, 42 pages.

California Herps (2020a) *Gartersnake;* San Francisco Gartersnake - *Thamnophis sirtalis tetrataenia. A Guide to the Amphibians and Reptiles of California.* www. californiaherps.com.

California Herps (2020b) Lizards. *A Guide to the Amphibians and Reptiles of California.* http://www.californiaherps.com/lizards/lizardspics.html.

California Native Plant Soc. (2017) California Native Plant Society Inventory of Rare and Endangered Plants of California; California Native Plant Society, Rare Plant Program. 2017; *Inventory of Rare and Endangered Plants of California* (online edition, v8-03 0.39).

Canadian Encyclopedia (nd) Wapiti. *The Canadian Encyclopedia.* Historica Canada, accessed December 23, 2016.

Carlton J.T., Thompson J.K., Schemel L.E., Nichols F.H. (1990) Remarkable invasion of San Francisco Bay (California, USA) by the Asian clam Potamocorbula amurensis. I. Introduction and dispersal. *Marine Ecology-*Progress Series 66: 81–94.

CDFW (2014) California red-legged frog named state amphibian. California Department of Fish and Wildlife. https://cdfgnews.wordpress.com/2014/07/15/california-red-legged-frog-named-state-amphibian/.

CDFW (nd-a) California Department of Fish and Wildlife. https://wildlife.ca.gov/Conservation/Mammals/Black-Bear/Population.

CDFW (nd-b) Keep Me Wild: Bobcat. California Department of Fish and Wildlife. https://wildlife.ca.gov/Keep-Me-Wild/Bobcat.

CDFW (nd-c) California condor. California Department of Fish and Wildlife. https://wildlife.ca.gov/Conservation/Birds/California-Condor.

CDFW (nd-d) Chinook salmon. https://wildlife.ca.gov/Conservation/Fishes/Chinook-Salmon.

CDFW (nd-e) Coho salmon. https://wildlife.ca.gov/Conservation/Fishes/Coho-Salmon.

Christian C.E. (2001) Consequences of a biological invasion reveal the importance of mutualism for plant communities. *Nature* 413: 635–639.

Conroy C.J., Rowe K.C., Rowe K.M.C., Kamath P.L., Aplin K.P., Hui L., James D.K., Moritz C., Patton J.L. (2012) Cryptic genetic diversity in *Rattus* of the San Francisco Bay region, California. *Biological Invasions* 15: 741–758.

Cronin M.A., Armstrup S.C., Garner E.R., Vyse G.W., Vyse E.R. (1991) Interspecific and intraspecific mitochondrial DNA variation in North American bears (Ursus). *Canadian Journal of Zoology* 69: 2985–2992.

Dobson A.P., Rodriguez J.P., Roberts W.M., Wilcove D.S. (1997) Geographic distribution of endangered species in the United States. *Science* 275: 550–553.

Dragoo J.W., Honeycutt R.L. (1997) Systematics of mustelid-like carnivores. *Journal of Mammalogy* 8: 426–443.

Economist (2012) Commercial whaling: Good whale hunting. *The Economist*, 4 March 2012.

Edford R.H., Wang X., Taylor B.E. (2009) Phylogenetic systematics of the North American fossil Caninae (Carnivora: Canidae). *Bulletin of the American Museum of Natural History* 325: 1–218, p. 146.

Edvenson J.C. (1994) Predator control and regulated killing: a biodiversity analysis. *UCLA Journal of Environmental Law and Policy* 13: 31–86.

Ellinger M. (2002) From the bottom up. *Bay Nature.* https://baynature.org/article/from-the-bottom-up/.

Ellis M. (2008) What native land snails live in the Bay Area? *Bay Nature.* https://baynature.org/article/what-native-land-snails-live-in-the-bay-area/#:~:text=Our%20wet%20and%20relatively%20mild,might%20as%20well%20eat%20them.

Feldhamer G.A., Thompson B.C., Chapman J.A. (2003) *Wild Mammals of North America: Biology, Management, and Conservation.* Baltimore, JHU Press. p. 683.

Fimrite P. (2019) New Point Reyes management plan riles up environmentalists—comment sought. *SF Chronicle.* https://www.sfchronicle.com/environment/article/New-Point-Reyes-management-plan-riles-up-14291769.php.

Flynn J.J., Finarelli J.A., Zehr S., Hsu J., Nedbal M.A. (2005) Molecular phylogeny of the Carnivora (Mammalia): Assessing the impact of increased sampling on resolving enigmatic relationships. *Systematic Biology* 54: 317–337.

Foster M.S., Schiel D.R. (1985) The ecology of giant kelp forests in California: a community profile. *US Fish and Wildlife Service Report* 85: 1–152.

Frost D.R. (2014) *Pseudacris regilla* (Baird and Girard, 1852). *Amphibian Species of the World: An Online Reference. Version 6.0.* American Museum of Natural History.

FWS (2007) California condor (*Gymnogyps californianus*). *U.S. Fish and Wildlife Service,* accessed September 28, 2020.

FWS (2020) California condor population information. US Fish and Wildlife Service. https://www.fws.gov/cno/es/CalCondor/Condor-population.html.

Goman M., Malamud-Roam F., Ingram B.L. (2008) Holocene environmental history and evolution of a tidal salt marsh in San Francisco Bay, California. *Journal of Coastal Research*: 24: 1126–1137.

Gong A.J. (2019) Voracious flowers of the tidepool. *Bay Nature.* https://baynature.org/2019/08/13/voracious-flowers-of-the-tidepool/.

Gould S.J. (1977). *Ontogeny and Phylogeny.* Belknap Press, Cambridge.

Green S. (2014) The last oyster. *Bay Nature.* https://baynature.org/article/last-oyster/.

Grenfell W.E., Jr (1974). *Food habits of the river otter in Suisun Marsh, Central California.* California State University, Sacramento. http://csus-dspace.calstate.edu/bitstream/handle/10211.9/1554/1974-Grenfell.pdf?sequence=1.

Groenendaal G.M. (1983) *Eucalyptus* helped solve a timber problem. *Proceedings of the Workshop on Eucalyptus in California,* pp. 1853–1880, Sacramento, CA.

Hairston N.G., Smith F.E., Slobodkin L.B. (1960) Community structure, population control and competition. *American Naturalist* 94: 421–425.

Hallmann C.A., Sorg M., Jongejans E., Siepel H., Hofland N., Schwan H., Stenmans W., Müller A., Sumser H., Hörren T., Goulson D., de Kroon H. (2017) More than 75 percent decline over 27 years in total flying insect biomass in protected areas. *PLoS ONE* 12(10): e0185809.

Hammerson G. (2008) *Rana draytonii. IUCN Red List of Threatened Species.* IUCN. 2008: e.T136113A4240307. doi:10.2305/IUCN.UK.2008.RLTS.T136113A4240307.en; http://www.fws.gov/sacramento/es/maps/CRF_fCH_FR_maps/crf_fCH_units.htm.

Harvell C.D., Montecino-Latorre D., Caldwell J.M., Burt J.M., Bosley K., Keller A., Heron S.F., Salomon A.K., Lee L., Pontier O., Pattengill-Semmens C., Gaydos J.K. (2019) Disease epidemic and a marine heat wave are associated with the continental-scale collapse of a pivotal predator (Pycnopodia helianthoides). *Science Advances* 5(1): eaau7042.

Haska C.L., Yarish C., Kraemer G., Blaschik N., Whitlatch R., Zhang H., Lin S. (2012) Bait worm packaging as a potential vector of invasive species. *Biological Invasions* 14: 481–493.

Hogan C.M. (2008) In California Giant Salamander: *Dicamptodon ensatus*; ed. N. Stromberg; globaltwitcher.com.

Ingram K.K., Gordon D.M. (2003) Genetic analysis of dispersal dynamics in an invading population of Argentine ants. *Ecology* 84: 2832–2842.

Jackman T.R., Wake D.B. 1994. Evolutionary and historical analysis of protein variation in the blotched forms of salamanders of the Ensatina complex (Amphibia: Plethodontidae). *Evolution* 48: 876–897.

Jaffe N., Eberl R., Bucholz J., Cohen C.S. (2019) Sea star wasting disease demography and etiology in the brooding sea star Leptasterias spp. *PLoS ONE* 14(11): e0225248.

Jennings M.R. (1983) *Masticophis lateralis* (Hallowel), Striped racer. Catalogue of American Amphibians and Reptiles. Society for the Study of Amphibians and Reptiles.

Jiang D., Klaus S., Zhang Y.-P., Hillis D.M., Li J.-T. (2019) Asymmetric biotic exchange across the Bering land bridge between Eurasia and North America. *National Science Review* 6: 739–745.

Johnson S., Gordon L. (2015) New maps reveal seafloor off San Francisco area. Sound Waves. US Geologic Survey. https://soundwaves.usgs.gov/2015/06/pubs.html.

Johnson W.E., O'Brien S.J. (1997) Phylogenetic reconstruction of the Felidae using 16S rRNA and NADH-5 mitochondrial genes. *Journal of Molecular Evolution* 44(suppl 1): S98–S116.

Jurek R.M. (1992) Non-native red foxes in California. Nongame Bird and Mammal Section Report. 92-04, California Department of Fish and Game.

Kenyon K.W. (1969) *The Sea Otter in the Eastern Pacific Ocean*. U.S. Bureau of Sport Fisheries and Wildlife, Washington, DC.

King J.L. (2004) The Current Distribution of the Introduced Fox Squirrel (*Sciurus niger*) in the Greater Los Angeles Metropolitan Area and its Behavioral Interaction with the Native Western Gray Squirrel (*Sciurus griseus*) Master's thesis, California State University, Los Angeles.

Klymkowsky M.W., Cooper M.M., Begovic E., Lymkowsky R. (2016) Sexual dimorphism. *Biology LibreTexts*. https://bio.libretexts.org/Bookshelves/Cell_and_Molecular_Biology/Book%3A_Biofundamentals_(Klymkowsky_and_Cooper)/04%3A_Social_evolution_and_sexual_selection/4.09%3A_Sexual_dimorphism, accessed September 27, 2020.

Krause W.J., Krause W.A. (2006) *The Opossum: Its Amazing Story*. Archived 2012-12-11 at the Wayback Machine. Department of Pathology and Anatomical Sciences, School of Medicine, University of Missouri, Columbia, MO. p. 39.

Kucera T. (1997) California Giant Salamander (Report). California Department of Fish and Game.

Kutcha S.R., Krakauer A.H., Sinervo B. (2008) Why does the yellow-eyed ensatina have yellow eyes? Batesian mimicry of Pacific newts (Genus *Taricha*) by the Salamander *Ensatina eschscholtzii xanthoptica*. *Evolution*. https://doi.org/10.1111/j.1558-5646. 2008.00338.x.

Kuchta S.R., Parks D.S., Lochridge Mueller R., Wake D.R. (2009) Closing the ring: historical biogeography of the salamander ring species Ensatina schscholtzii. *Journal of Biogeography* 36: 982–995.

Lang G. (2020) The smallest stars have gone out. *Bay Nature*. https://baynature.org/2020/02/11/the-smallest-stars-have-gone-out/.

Lantz D.E. (1909) The brown rat in the United States. US Department of Agricultural Biological Survey Bulletin 33: 1–54.

Larsen D.N. (1984). Feeding habits of river otters in coastal southeastern Alaska. *Journal of Wildlife Management* 48: 1446–1452.

Leopold A. (1949) Thinking like a mountain. In edited by A. Sand, *County Almanac: And Sketches Here and There*. Oxford University Press.

Liebhold A.M., Kean J.M. (2019) Eradication and containment of non-native forest insects: successes and failures. *Journal of Pest Science* 92: 83–91.

Lister B.C., Garcia A. (2018) Climate-driven declines in arthropod abundance restructure a rainforest food web. *Proceedings of the National Academy of Sciences of the United States of America* 115: E10397–E10406.

Litalien R. (2004) *Champlain: The Birth of French America*. Montreal, Quebec, McGill-Queen's Press. pp. 312–314.

Maffe W.A. (2000) A note on invertebrate populations of the San Francisco Estuary. In: Goals Project. 2000. *Baylands Ecosystem Species and Community Profiles: Life Histories and Environmental Requirements of Key Plants, Fish and Wildlife*. Prepared by the San Francisco Bay Area Wetlands Ecosystem Goals Project. P.R. Olofson, editor. San Francisco Bay Regional Water Quality Control Board, Oakland, CA, pp. 184–192.

Martin G. (2011) The Middle Way. *Bay Nature*, July-September 2011. https://baynature.org/article/the-middle-way/, accessed September 27, 2020.

Martín-Durán J.M., Passamaneck Y.J., Martindale Mark Q., Hejnol Andreas (2016) The developmental basis for the recurrent evolution of deuterostomy and protostomy. *Nature Ecology & Evolution* 1: 0005.

Mathiasson M.E., Rehan S.M. (2019) Status changes in the wild bees of Northeastern North America over 125 years. https://doi.org/10.1111/ icad.12347.

Mayer J.J., Brisbin I.L. Jr (2008) *Wild Pigs in the United States: Their History, Comparative Morphology, and Current Status*. Athens, Georgia, University of Georgia Press. p. 20.

McKercher L., Gregoire D.R. (2020) *Lithobates catesbeianus* (Shaw, 1802): U.S. Geological Survey, Nonindigenous Aquatic Species Database, Gainesville, FL.

Melquist W.E., Dronkert A.E. (1987) River otter. In *Wild Furbearer Management and Conservation in North America*, edited by Novak M., Baker J.A., Obbard M.E., Malloch B. Ontario Ministry of Natural Resources, Toronto, Canada. pp. 626–641.

Miller C.R., Waits L.P. (2006) Phylogeography and mitochondrial diversity of extirpated brown bear (*Ursus arctos*) populations in the contiguous United States and Mexico, *Molecular Ecology* 15: 4477–4485.

Moffett M.W. (2012) Supercolonies of billions in an invasive ant: what is a society? *Behavioral Ecology* 23: 925–933.

Moritz C., Schneider C.J., Wake D.B. (1992) Evolutionary relationships within the *Ensatina eschscholtzii* complex confirm the ring species interpretation. *Systematic Biology* 41: 273–291.

Morris R.H., Abbott D.P., Haderie E.C. (1980) *Intertidal Invertebrates of California*. Stanford University Press. Palo Alto, CA.

Murray W. (2004). *Elsevier's Dictionary of Reptiles*. Amsterdam, Elsevier. p. 122.

National Ocean Service (2013) What lives in a kelp forest: Kelp forests provide habitat for a variety of invertebrates, fish, marine mammals, and birds NOAA. Updated 11 January 2013. https://oceanservice.noaa.gov/facts/kelp.html#:~:text=Among%20the %20many%20mammals%20and,cold%2C%20nutrient%2Drich%20waters, accessed September 27, 2020.

National Ocean Service (2020) Kelp forests: a description. https://sanctuaries.noaa.gov/ visit/ecosystems/kelpdesc.html, accessed September 27, 2020.

Neilson S. (2019) More bad buzz for bees: Record number of honeybee colonies died last winter. KQED. https://www.npr.org/sections/thesalt/2019/06/19/733761393/more-bad-buzz-for-bees-record-numbers-of-honey-bee-colonies-died-last-winter#:~:text= Bee%20decline%20has%20many%20causes,systems%20and%20can%20kill %20them, accessed September 30, 2020.

NOAA (2020) U.S. Department of Commerce National Oceanic & Atmospheric Administration National Environmental Satellite, Data, and Information Service; www.ncdc.noaa.gov, accessed September 24, 2020.

Norwak R.M. (1999) *Walker's Mammals of the World*. Baltimore, JHU Press. p. 1521.

NPS (2018) Marine invertebrates. National Park Service. https://www.nps.gov/goga/learn/nature/marine-invertebrates.htm, accessed September 28, 2020.

Oksanen L., Fretwell S.D., Arruda J., Niemala P. (1981) Exploitation ecosystems in gradients of primary productivity. *American Naturalist* 118: 240–261.

Ort B.S., Cohen S., Boyer K.E., Wyllie-Echeverria S. (2012) Population structure and genetic diversity among eelgrass (*Zosteria marina*) beds and depths in San Francisco Bay, *Journal of Heredity* 103: 533–546.

Ortiz J.L., Muchlinski A.E. (2014) Urban/suburban habitat use by a native and invasive tree squirrel. *Bulletin of the Southern California Academy of Sciences* 113: 116.

Parr I. (2019) Naturally, 2019 Closes with Thousands of 10-Inch Pulsing "Penis Fish" Stranded on a California Beach. *Bay Nature*. https://baynature.org/2019/12/10/naturally-2019-closes-with-thousands-of-10-inch-pulsing-penis-fish-stranded-on-a-california-beach/, accessed September 28, 2020.

Paton T.A., Baker A.J., Groth J.G., Barrowclough G.F. (2003) RAG-1 sequences resolve phylogenetic relationships within charadriiform birds. *Molecular Phylogenetics and Evolution* 29: 268–278.

Pecon-Slattery J., O'Brien S.J. (1998) Patterns of Y and X chromosome DNA sequence divergence during the Felidae radiation. *Genetics* 148: 1245–1255.

Pereir R.L., Wake D.B. (2009) Genetic leakage after adaptive and nonadaptive divergence in the *Ensatina eschscholtzii* ring species. *Evolution* 63(9): 2288–2301.

Pons J.-M., Hassanin A., Crochet P.-A. (2005) Phylogenetic relationships within the Laridae (Charadriiformes: Aves) inferred from mitochondrial markers. *Molecular Phylogenetics and Evolution* 37: 686–699.

Poppe K.L., Rybcyk J.M. (2018) Carbon sequestration in a Pacific Northwest eelgrass (*Zostera marina*) meadow. *Northwest Science* 92(2): 80–91.

Poulter G. (2010) *Becoming Native in a Foreign Land: Sport, Visual Culture, and Identity in Montreal, 1840-85*. Vancouver, UBC Press. p. 33.

Presidio Trust Website. https://www.presidio.gov/presidio-trust/planning/coyotes-in-the-presidio, accessed September 25, 2020

Reynolds J.W. (2016) Earthworms (Oligochaeta: Acanthodrilidae, Lumbricidae, Megascolecidae, Ocnerodrilidae and Sparganophilidae) in the Central California Foothills and Coastal Mountains Ecoregion (6), USA. *Megadrilogica* 21: 73–78.

Rice D.W. (1998). *Marine Mammals of the World. Systematics and Distribution*. Special Publication Number 4. The Society for Marine Mammalogy, Lawrence, KA.

Ripple W.J., Estes J.A., Beschta R.L., Wilmers C.C., Ritchie E.G., Hebblewhite M., Berger J., Elmhagen B., Letnic M., Nelson M.P., Schmitz O.J., Smith D.W., Wallach A.D., Wirsing A.J. (2014) Status and ecological effects of the world's largest carnivores, *Science* 343, 1241484

Robbins C.S., Brunn B., Zim H.S. (1983) *Birds of North America; A Guide to Field Identification*, Golden Press, New York, NY.

Rodriguez O.R. (2020) Roaming mountain lion caught in downtown San Francisco, AP News June 18, 2020; https://apnews.com/article/fd333357dbe7b8d3851b0418b9ea3ecf.

Roth B., Sadeghian P. (2006) *Checklist of the Land Snails and Slugs of California*. 2nd ed. Santa Barbara, California, Santa Barbara Museum of Natural History.

Rubenstein S. (2020) More than a dozen California condors missing after wildfire destroys their Big Sur sanctuary. San Francisco Chronicle, August 26, 2020; https://www.sfchronicle.com/california-wildfires/article/Fires-destroy-Big-Sur-condor-sanctuary-15516997.php.

Rudnick D.A., Kieb K., Grimmer K.F., Resh V.H. (2003) Patterns and processes of biological invasion: The Chinese mitten crab in San Francisco Bay. *Basic and Applied Ecology* 4: 249–262.

Rychel A.L., Smith S.E., Shimamoto H.T., Swalla B.J. (2006) Evolution and development of the chordates: collagen and pharyngeal cartilage. *Molecular Biology and Evolution* 23(3): 541–549.

Sacks B.N., Moore M., Statham M.J., Wittmer H.U. (2011) A restricted hybrid zone between native and introduced red fox *Vulpes vulpes* populations suggests reproductive barriers and competitive exclusion. *Molecular Ecology* 20: 326–341.

Sahagún L. (2020) Southern California mountain lions get temporary endangered species status. *Los Angeles Times*. https://www.latimes.com/environment/story/2020-04-16/state-panel-studying-threatened-species-protection-for-southern-california-cougars.

San Diego Zoo News report, undated California Condor – San Diego Zoo Animals & Plants. animals.sandiegozoo.org

Santos, R.L. (1997) *The Eucalyptus of California*, Alley-Cass, Denair, CA.

SelecTree (2020) SelecTree. California buckeye. *Aesculus californica* Tree Record. 1995–2020. https://selectree.calpoly.edu/tree-detail/aesculus-californica.

Shaffer, H.B., Pauly, G.B., Oliver, J.C., Trenham, P.C. (2004). The molecular phylogenetics of endangerment: cryptic variation and historical phylogeography of the California tiger salamander, *Ambystoma californiense. Molecular Ecology* **13** (10): 3033–3049.

Shelley R.M. (2002) Annotated Checklist of the Millipeds of California (Arthropoda: Diplopoda). *Monographs of the Western North American Naturalist* 1: 90–115.

Simmons E. (2019) The importance of having insects. *Bay Nature*. May 2019.

Stebbins R.C. (1949) Speciation in salamanders of the plethodontid genus *Ensatina. University of California Publications in Zoology* 48: 377–526.

Stebbins R.C. (1954) Natural history of the salamanders of the plethodontid genus *Ensatina. University of California Publications in Zoology* 54: 47–124.

Stebbins R.C. (1959) *Reptiles and Amphibians of the San Francisco Bay Region.* University of California Press, Berkeley and Los Angeles, CA.

Stebbins R.C. (1972) *California Amphibians and Reptiles.* University of California Press, Berkeley, Los Angeles, and London.

Stebbins R.C. (2003) *A Field Guide to Western Reptiles and Amphibians*, 3rd ed. The Peterson Field Guide Series. Houghton Mifflin Company, Boston, MA and New York, NY.

Striedter G.F., Avise J.C., Ayala F.J. (2013) *In the Light of Evolution: Volume VI: Brain and Behavior.* Washington, DC, National Academies Press.

Sunset (1995) *Sunset Western Garden Book*; 40th anniversary edition. June 1995. Sunset Publishing Corporation, Menlo Park, CA 94025; pp 28–31.

Swaim K.E. (1994) *Aspects of the Ecology of the Alameda Whipsnake (Masticophis lateralis euryxanthus).* Unpublished Master's Thesis, California State University, Hayward, 140 pp.

Swaim K.E., McGinnis S.M. (1992) Habitat associations of the Alameda whipsnake. *Transactions of the Western Section of The Wildlife Society* 28: 107–111.

Swensrud A., Platt M. (2007) *South San Francisco Bay Invertebrate Guide.* Marine Science Institute. https://sfmsi.files.wordpress.com/2014/03/2007-benthic-guide-cover-and-contents.pdf.

Tan A.-M., Wake D.B. (1995) MtDNA phylogeography of the California newt, Taricha torosa (Caudata, Salamandridae). *Molecular Phylogenetics and Evolution* 4: 383–394.

Tighe D. (2019) Meet the Bay's incredible swimming worms. *Bay Nature*. https://baynature.org/2019/08/06/meet-the-bays-incredible-swimming-worms/.

Ueda K.-I., Agarwal R.G. (nd) California sea slugs – Nudibranchs (and other marine Heterobranchia) of California. iNaturalist. https://www.inaturalist.org/guides/40.

USGS 2780 (nd) https://nas.er.usgs.gov/queries/FactSheet.aspx?SpeciesID=2780.

van Tuinen M., Waterhouse D.M., Dyke G.J. (2004) Avian molecular systematics on the rebound: a fresh look at modern shorebird phylogenetic relationships. *Journal of Avian Biology*. 35: 191–194.

Vetter R.S. (2000) Myth: Idiopathic wounds are often due to brown recluse or other spider bites throughout the United States. *Western Journal of Medicine* 173: 357–358.

Wake D.B. (1997) Incipient species formation in salamanders of the Ensatina complex. *Proceedings of the National Academy of Sciences of the United States of America* 94: 7761–7767.

Watkins B. (nd) California marine flatworms. California Diving. https://cadivingnews.com/california-marine-flatworms/.

Wallach A.D., Johnson C.N., Ritchie E.G., O'Neill A.J. (2010) Predator control promotes invasive dominated ecological states. *Ecology Letters* 13: 1008–1018.

Williams B.L., Brodie E.D. III (2003) Coevolution deadly toxins and predator resistance: self-assessment of resistance by garter snakes leads to behavioral rejection of toxic newt prey. *Herpetologica* 59: 155–163.

Wilson D.E., Mittermeier R.A. (eds) (2009) *Handbook of the Mammals of the World, Volume 1: Carnivora*. Barcelona, Lynx Ediciones, pp. 50–658.

Wolf K.M. (2015) Management of blue gum eucalyptus in California requires region-specific consideration. *California Agriculture* 70: 39–47.

Woodward S.L., Quinn J.A. (2011) *Encyclopedia of Invasive Species: From Africanized Honey Bees to Zebra Mussels*. Santa Barbara, California, ABC-CLIO.

Wozencraft W.C. (2005) Order Carnivora. In edited by Wilson D.E., Reeder D.M. *Mammal Species of the World: A Taxonomic and Geographic Reference* (3rd ed.). Johns Hopkins University Press, pp. 624–628.

9 Restoring the Bay

THE BAY IS NOT WHAT IT ONCE WAS

The Bay was millions of years in the making. Plate tectonics had created the structure we see today perhaps 2 million years ago. Since then, during several glacial and interglacial periods the Bay has filled with water and emptied a number of times (Barnard et al., 2013). Today water flows into the Sacramento and San Joaquin Rivers through the Carquinez Strait and on into the Bay and out the Golden Gate into the Pacific Ocean. That basic flow was established about 600,000 years ago. The most recent flooding of the estuary occurred 10,000–11,000 years ago.

Over 150 years of human exploitation have caused the Bay to deteriorate. More than 85% of the tidal marsh has been lost to filling and building. The boundaries of the Bay have changed significantly. Rivers and streams and even the Bay have been filled with silt and pollution. More modern efforts have attempted to contain rivers within their beds to reduce the threats of flooding. Pollutants have been swept into the Bay from sewage systems (or simple raw sewage), industrial processes, and agricultural activities.

Climate change is already upon us, and the Earth is warming at an increasing rate. As the ice caps melt, sea levels rise, rain patterns change, the temperature of bodies of water is increasing, freshwater becomes less available, and changes occur to many other systems. All of this change will stress the plants and animals around the world. Not surprisingly, those changes are also being felt in the Bay Area.

Global warming and rising ocean levels are likely to threaten even more wetlands. The Bay is silting up by mining and other practices. Pollution began with mining, but has expanded to include agricultural runoff, sewage dumping or releases, filling of the shoreline, damage to wetlands, and more. The South Bay shorelines provide a good example of how human activities changed the shorelines of the Bay. In the early part of the 20th century, much of that part of the Bay was transformed into salt pond evaporators. These were formed by building levees to impound portions of the salt marsh. In addition, parts of the shore were built on. Moffett Field near Mountainview was built on fill on former wetlands.

The Bay cannot go back to what it was 10,000 years ago or even 200 years ago. More than 7.5 million people live and work here. The problem will get worse as more people more here, housing expands upward, and sea levels rise. Still, the wetlands are critical habitat. Each year more than 500,000 shorebirds and 700,000 waterfowl use the mudflats and salt ponds to rest and feed as they migrate along the Pacific Flyway.

Balancing the competing needs of those birds that need habitats with open water or tidal flats with others that need the tidal marshlands is a challenge. And beyond the needs of birds and other species are the needs of the growing human population in the Bay Area. Some balance needs to be struck so that the birds and other living organisms can thrive and people can live here and enjoy the natural beauty of the Bay. Both goals can be at least partially met.

Fortunately, in the last few decades, efforts have begun to maintain and even restore the Bay to a more natural state. Several new laws, such as the Clean Water Act of 1972, combined with California's Porter Cologne Act of 1969 and the McAteer-Petris Act of 1965 and others, brought the force of law to the conservation side. These new efforts focus on many aspects, such as controlling pollution and contamination and restoring the bay shore. Stralberg et al. (2011) examined the loss by using established models. Their hybrid approach combined elevation data with estimates of the accumulation of organic matter and sediment deposition in 15 Bay subregions. They concluded that bay lands and mid-marsh areas could be restored with a reasonable amount of material. However, significantly more material would be needed to save those same wetlands over 100 years.

The restorations come in many forms: restoring salt ponds to more natural state, restoring wetlands along the shore and the rivers feeding the Delta, transforming former military bases, accommodating wildlife, eliminating invasive species, and defending against sea level rise. Restoration work is highly desired, but it requires careful planning and execution to ensure that the unintended consequences are minimized. Fortunately, there are some hopeful signs.

THE MAIN BAY

The Main Bay is the deepest part of the Bay and has the strongest tidal currents and the most coarse sediment. The currents are stronger in the western area, and much of the sediment is sand. East of Angel Island and Alcatraz Island, the sediment is more mud mixed with sand.

One of the most important issues for the Main Bay is dredging. The Bay has 85 miles of navigable waterways that require dredging to keep them open. Dredging necessarily churns up a cloud of sediment that can move in different directions. Those directions are a function of the sediment and water characteristics, and understanding their interactions can help minimize the detrimental effects of dredging on marine life. Capello et al. (2010) studied those factors, including the winds and waves, turbidity, characteristics of the sediments, and more.

Dredging has many adverse effects on marine life (Fraser et al., 2017). It increases sedimentation and reduces light levels in the area. Marine organisms that depend on photosynthesis can be harmed by the loss of light. The additional suspended material might also interfere with feeding and other activities. For example, oysters are sensitive to the effects of dredging (Wilber and Clarke, 2010). First, the oysters can be physically covered by material settling after dredging, and second, dredging dramatically increases sediment levels, which can interfere with the oyster's filter feeding. Finally, oyster larvae need a clean hard surface so they can attach to it. Even small amounts of loose sediment can disrupt that attachment and seriously damage the oyster population. In most areas, new regulations limit the amount of dredging around oysters.

For many years, dredged material was considered waste to be disposed of. However, the San Francisco Estuary Partnership said, "Like freshwater, sediment is a precious resource that is essential for keeping the Estuary healthy" (San Francisco Estuary Partnership, 2015). Sediment restores shorelines, beaches, and wetlands, and the balance of material washed further downstream and that newly delivered is critical to maintaining the environment where land meets water (Milligan and Holmes, 2017). Unfortunately, disruption of the balance of loss and gain of sediment has large-scale effects, and human interventions have often disrupted the balance. For example, the Bay is a major shipping port, and dredging is needed to ensure that a channel is open for ships to safely enter and exit the Bay. Dams and other structures upstream also interrupt the flow of water and sediment.

Contamination of sediment is a serious issue. Since the 1940s, large amounts of sediment have been dredged from shipping channels in the Bay and dumped, along with waste from oil refineries and steel mills, various chemicals, and even low-level radioactive waste, into the ocean near the Farallon Islands (Chin and Ota, 2001). One common contaminant is polychlorinated biphenyls (PCBs) (Yee and Wong, 2019). In 2008, the San Francisco Bay Regional Water Quality Control Board established a maximum daily load for PCBs in the Bay, and a plan to limit PCB concentrations reduces addition of PCBs from the storm drains mainly, controls dredging, and advises consumers of PCB risks in fish from the Bay. These and other efforts have helped to greatly reduce PCB levels in the Bay. They also guide the disposition of dredged sediment from the Bay.

Fortunately, more recently, sediment has come to be valued, and a number of studies have sought to better understand the movement of sediment through the whole Bay system. The Golden Gate is the link between the Bay and the Pacific Ocean and the next-to-last stop for sediment that began its journey in the Sierras and will wind up on the open coast. Understanding sediment transport through that strait is important for understanding the overall movement of material through the Bay system. The amount of sediment (mostly as sand) lost from the Bay over the last 50 years is estimated to be more than 150 million m^3. Loss of that material has detrimental effects on the beaches and wetlands. Modeling by Elias and Hansen (2013) showed that the tides control the movement of sediment at the Golden Gate. Overall, the flow out of the Bay dominates the flow inward, and so, material collects in an ebb-tide delta outside the Gate and more so on the deeper north side of the channel. Ocean Beach on the San Francisco shore is one result of the deposition of this sediment. Barnard et al. (2013) determined the transport paths from the source to the sinks of sand on the outer coast. They found that human intervention is a major factor in the loss of sediment in the Bay.

Climate change has brought the threat of sea level rise and increased storm surge to coastal areas, including the Bay. The storms, in particular, cause enormous damage. Salt marshes and other natural barriers are effective defenses against these threats, and efforts are being made to improve those barriers. To protect against storm damage, shorelines have been raised. Global warming will make this work even more urgent. The spoils from dredging are useful material for this purpose. The salt marshes and mud flats are outstanding barriers to flooding. Of course, building them up with dredged material is often only a temporary fix. The transferred material will soon enough seek equilibrium by refilling the dredged channel (Ganju, 2019). The addition of new material slows this process, but an understanding of the overall movement of sediment through the whole Bay system is needed.

Of the material dredged from the shipping channels in the Bay, only about 20% is allowed to be deposited in open-water in-Bay placements. The rest is transported to a deep ocean dump site off-shore. Current law also requires that 40% of the material be put to good use in mud and salt flats. Sea level rise will exacerbate the problem and so require even more dredged material to keep those areas above the water. The location of the placement of the material was a critical factor (Bever et al., 2014). The placements in San Pablo Bay were not effective in adding material to the surrounding mudflats, but those in the South Bay were. The factors acting on the sediment are complicated. The interaction of tides, winds, waves, and sediments must be modeled so that appropriate decisions can be made about placements. Their model has some limitations, but gives a good approximation, especially when comparing different scenarios. They conclude that placements in the open water can be used effectively to augment mudflats.

WETLANDS AND SALT PONDS

Wetlands are places where shallow water covers the land. They constitute only 6% of the Earth's surface, but they provide food, water, and refuge for enormous numbers of humans and animals (Greb et al., 2006). The first land plants likely appeared in a wet environment, and many fossils were found in former wetlands. Wetlands today are among the most productive ecosystems on Earth, but they are threatened and disappearing rapidly. They provide habitat and food for diverse species, retain water and recharge groundwater, reduce flooding, prevent erosion, control sedimentation, filter water, and cycle nutrients. Greb et al. (2006) is a tour-de-force description of the importance of wetlands today and throughout Earth's history.

Wetlands have a secret weapon in protecting the land against sea level rise (Gies, 2018). They accrete sediment, and plants grow, die, and build up over time. Thus, the wetlands also grow in height. Their growth is especially important in the Bay Area. Sea levels are rising, and the Bay has an extensive shoreline, including the Delta with its vulnerable islands.

Olympia oysters (*Ostrea lurida*) are an excellent model for testing the adaptation to changes in the local environment and local adaptation with applications to marine restoration. Bible and Sanford (2016) used them as a model system to examine their ability to adapt to conditions of low salinity, which they are likely to experience as our climate changes in the near future. They had two tests. First, they took oysters from three locations in the Bay and held them in laboratory conditions for a while before returning them to different wild locations. Oysters that survived better were returned to their home location. In the second, they took oysters from the two sites in the Bay and also from Tomales Bay, north of San Francisco Bay. They held them under laboratory conditions and then subjected them to conditions of low salinity. They found that oysters from areas of low salinity survived the best. They concluded that local adaptability is likely a key factor in planning how to best restore species in the natural environment.

The South Bay Salt Pond Restoration Project is the largest wetland restoration project on the West Coast (Shellenbarger et al., 2013). The hope is to restore 61 km^2 that was previously owned by the corporation Cargill and used for salt production.

By 2060, the South Bay Salt Pond Restoration Project seeks to restore up to 54 km^2 of tidal wetlands from the salt ponds. For this purpose, they will need to capture the estuarine mud. Is there enough mud to go around? A model of the Bay mud (Brew and Williams, 2010) was developed to examine multiple scenarios involving no intervention or various restoration efforts. The model predicted that mud will not be the limiting factor in the restoration. However, the effects of sea level rise are not completely understood.

Restoring the marsh lands is difficult enough, but this project has additional problems. Historical mining practices that resulted in contamination of the area by mercury. Mining for gold in the Sierra Nevada and for mercury in the Coastal Ranges contributed to the pollution. Now the mercury is contained in the sediments in the Bay. As long as the mercury in the sediments are not disturbed, they pose no problem to the environment. However, the efforts to restore the salt ponds will disturb the sediments. Once it is liberated, it has a very good chance of getting into the aquatic food web.

Some of the restoration work began in the Alviso Slough in the South Bay. Even with the best efforts, the work disturbed the mercury-laden sediments. Foxgrover et al. (2019) developed a new method for determining mercury levels and found that 52 (=/– 3) kg of mercury was remobilized in the 6 years that the project has been under way. The great majority of the erosion occurred in the winter months.

Another complication for the project is that the area is used by thousands of birds for over-wintering and as rest areas during migration. Thus, the project calls for the construction of islands and berms and other measures to maintain water salinity and depth so the effects of the restoration will be minimal on the birds. As part of this work, De La Cruz et al. (2018) evaluated the number of water birds at the site for October to April for 13 years and determined the types of ponds that were most favored. These data were used to guide the restoration work.

Water bird numbers have declined throughout the world due to many factors, including loss of habitat, loss of prey, pollution, overharvest, and invasive species. Unfortunately, if not done correctly, wetlands restoration can add to those threats and accelerate the loss of birds. Even if attractive habitat is established, it might not be enough to bring back the birds. Hartman et al. (2019) built new islands and changed the substrate to make it attractive. They also used sound recordings of breeding colonies to entice new birds to begin to use the restored wetland areas.

They focused on the Caspian tern (*Hydroprogne caspia*). Terns have lost much of their traditional nesting habitat. In the Bay, they have declined in numbers from 1000 breeding pairs in 1982 to 830 in 2009 (Hartman et al., 2019). The number and size of the colonies have also declined. However, terns are not as programmed as other birds to return to the location of their birth, and they respond to social attraction measures, such as decoys and recordings of colony sounds.

These efforts were successful at two locations. Over 500 breeding pairs were counted, and over 3 years, they estimated that their two colonies yielded over 1300 nests and more than 530 fledglings. These experiments demonstrate the power of these methods for restoration.

The Napa River Salt Marsh Restoration Project is another effort to bring back tidal marsh. This project seeks to restore 10,000 acres of former salt ponds in the North Bay. The original marsh included 25,000 acres, but 64% of it was taken over by farming and development. In 1994, Cargill Salt Company sold 9800 acres to California. The project requires the removal of "bittern." Bittern results when saltwater is evaporated and contains table salt and other ions, such as magnesium, calcium, and potassium. In addition, the water flow must be improved so it will support bird and other populations.

The Cullinan Ranch Tidal Marsh Restoration Project will restore 1549 acres of tidal wetlands in the San Pablo Bay National Wildlife Refuge. This area was diked in the late 19th century and transformed into farm land. In 1991, it became part of the Refuge. Now the work will open levees so that sedimentation can restore the marshlands while protecting existing infrastructure.

As tidal marsh is restored to its natural state, they still need to be managed to ensure that they are optimal for all species and for the numbers of organisms that use them. Different birds need different habitats. Diving ducks need deeper water. Small shorebirds need shallow water. They all need areas where the can both rest and roost.

DELTA RESTORATION

The Delta comprises a maze of interconnected channels and islands. The island formed long ago and are protected by over 1700 km of levees. The interiors of the islands have eroded and been compacted mostly by agricultural used over the last 150 years. The levees have been weaken by many factors, such as the rise of seawater due to climate change, flooding, subsidence, mammal burrows, and even earthquakes. Loss of the levees can result in loss of agricultural assets and drinking water as saltwater drowns the islands. Most of the islands are in private hands. Repair and maintenance of the levees is a critical issue.

The state of California has begun managing floodplains. To do this, they are breaching levees so that floodwaters have somewhere to go. The idea is that letting the water flood in a planned area will prevent flooding in other areas. This protects human infrastructure and agriculture. It also is good for fish and other animals. Finally, use of this concept will be even more important in the future when climate change brings more extreme events. Warmer temperatures will also cause more winter precipitation to fall as rain, rather than snow, in the Sierra Nevada. However, the floodplains will soak up the water and help to recharge aquifers that have been depleted to provide water during recent severe droughts.

Over the last couple of centuries, considerable efforts were made to control rivers and limit flooding. Rivers were channeled and dammed. In more recent years, the importance of rivers and floodplains has been realized. Rivers are much more healthy when they are connected to a floodplain. Matella and Merenlender (2014) examined the association of river flow and floodplain area. More specifically, they looked at how splittail (*Pogonichthys macrolepidotus*), Chinook salmon (*Oncorhynchus tshawytscha*), and their food resources in the San Joaquin River would be influenced by two scenarios of restoration and climate change. They found that the frequency of floods sufficient to ensure splittail and salmon rearing was relatively rate. The project that only 4–17% of the years of the rest of the 21st century will produce enough waterflow and flooding to assure habitat restoration.

POLLUTION IN THE BAY

For many years, the Bay was an easy way to get rid of unwanted items. During the Gold Rush days, old buildings and rotting ships were thrown into the Bay as fill to expand the amount of land available for development. Since then, industrial waste, mine tailings, agricultural runoff, raw sewage, ordinary trash, used oil, and much more has wound up in the Bay. It is not just sewage that reaches the Bay. Chemicals, bacteria, pesticides, motor oil, heavy metals, and dirt also reach the Bay. Dirt and sand can be beneficial by providing additional sediment that forms homes for many aquatic organisms. Even the sediment can be polluted with mercury from long-closed mines in nearby regions. Fortunately, much of those items are now much better controlled. Here we will review a couple of significant types of pollution.

MERCURY

Mercury contamination is a serious problem, and estuaries are often heavily contaminated due to the large numbers of people who tend to live near them and the activities that take place there. In estuaries, methylmercury is the main contaminant. Estuaries are highly productive, and it is easy for the mercury to enter the food chain. As one of the largest urban area in the United States, San Francisco Bay is a good example. The mixture of methyl contaminants is unusual and dominated by the legacy of mining from the 19th century. Much of it is cinnabar and other sulfide-complexed forms. There are no coal-fired power plants in the Bay Area, but interestingly, a significant contributor to Bay Area contamination is mercury from other parts of the world, such as Asia. Other factors include the disruption of the historic wetlands along the Bay and the reduction of incoming sediment.

The gold rush is over, but its legacy lives on. Many abandoned mines continue to leak the mercury that was used to process gold. Most of that leakage flows into the Delta and then the Bay. However, other mining operations are also problems. The New Almaden Mine in Santa Clara County was the largest mercury mining operation in North America, and its legacy has contaminated the South Bay.

The Guadalupe River Watershed Mercury TMDL determined that the main source of mercury contamination was the New Almaden Mining District. For more than 100 years, the mining waste left over from the Mine contaminated the Guadalupe River and local creek beds that flow to the Bay. Other sources include atmospheric deposition, soil erosion from non-mining areas, urban storm runoff, and seepage from landfills. The TMDL is working to reduce contamination from these sources. For example, since 2008, the state of California has been implementing a restoration project. The Hacienda and Deep Gulch Remediation Project, approved by Santa Clara County in 2010, seeks to eliminate mercury contamination from the largest mercury ore smelting plant in the area.

However, the Bay is carefully monitored now, and efforts are under way to reduce methyl contamination. Davis et al. (2012) published a comprehensive review of methymercury pollution in the Bay and suggested a number of ways to reduce levels of the contaminant. Advisories discourage people from consuming fish, particularly striped and largemouth bass, from the Bay. These accumulations also affect other species and indicate the depth of the problem.

Unfortunately, Davis et al. (2012) suggest that this situation will require decades of work to fix. Some of the elements of an overall program to reduce mercury contamination include improved mining practices and measures to reduce urban runoff into the Bay. Attempts to control the internal net production of methylmercury are also valuable. Efforts to control contamination in upstream reservoirs and ponds may help. The contamination in the Bay sediments is heavy, and there are few options for cleaning it up. Still other factors may be difficult to control to reduce bioaccumulation. These include climate change, sea level rise, and introduced species.

SEWAGE

In 2017, the Bay Area broke out of a drought with very heavy storms. The rain was welcome, but it had one downside. More than 120 sewage overflows occurred within 3 weeks. More worrisome, 85 of those overflows reached the Bay. More than 9 million gallons of partially treated material from the West Contra Coastal Sanitary District wound up in a marsh connected to the Bay. The city of Vallejo released 2 million gallons and the city of San Mateo released 260,000 gallons. Finally, the East Bay Municipal Utility District (EBMUD) spilled 5 million gallons of raw sewage into the Bay. The only small positive aspect of these spills was that the material was greatly diluted. Nevertheless, allowing storm runoff or sewage into the Bay is not healthy.

Although progress has been made in controlling sewage spills into the Bay, the municipal sewer systems are still not capable of controlling and treating waste especially during the rainy season. The rain gets into the sewer system from leaky pipes, and the additional water can overwhelm treatment facilities. Many of those facilities and pipelines are old and fragile. For example, in August 2020, a power failure caused treatment facilities to shut down, and 50,000 gallons of untreated sewage flowed into the Bay in Oakland and Alameda (Morris, 2020).

Each winter, as the rains begin, millions of gallons of raw and undertreated sewage flow into the Bay. Raw sewage is a significant health concern. It contains bacteria, viruses, and other disease-causing organisms. The many people who love water sports may come into contact with those pathogens when they are on the Bay. Sewage also lowers oxygen levels, which puts fish, seals, other sea creatures, and plant life under stress. In addition to pathogens, sewage changes the nutrient supply in the water. Autotrophs, such as algae, may grow out of control, especially when nitrogen and phosphorus levels are increased. Once dead, the algae settle to the bottom, and decomposition reactions can use up much of the available oxygen so that some species must leave the area.

Since the 1970s, the Bay has become much cleaner. Pollution controls and better handling of sewage have clearly helped. Still, more needs to be done to prevent the runoffs, such as in 2017. Replacing leaky lateral sewage pipes will prevent rain from getting into the sewage system. The EBMUD has started an aggressive plan to upgrade its system and to encourage homeowners to replace laterals.

TRANSFORMING MILITARY BASES

The military has been a part of the Bay Area since the Spanish established the Presidio of San Francisco in 1776. To defend San Francisco and the port, the US military continued to maintain forts around the Golden Gate until the mid-1990s when the land was turned over to the Department of the Interior. Much of it became the Golden Gate National Recreation Area. The best part of the transfer was that a large amount of undeveloped coastal land was available to everyone.

Unfortunately, not all of the land transfers have been as successful. A number of the active installations were heavily contaminated with petroleum products, toxic chemicals, and radioactive material. Cleanup has been long and expensive at bases, such as Hunters Point (Perelo, 2010; Cho et al., 2012) and Alameda Naval Air Station (Love et al., 2003).

WILDLIFE IN AN URBAN ENVIRONMENT

Most wild animals typically avoid urbanized areas, but some have adapted in amazing ways. In an extreme case, a coyote was captured in San Francisco, DNA tested, and released (Sacks et al., 2006). The animal's DNA sequences showed that it belonged to the population in Marin County and, thus, the coyote must have crossed the Golden Gate Bridge. More interestingly, the bridge may have become a mechanism to allow two historically distinct populations of coyotes to interbreed. Coyotes seem to have moved into San Francisco in about 2004, and now, they are well established in many neighborhoods and terrorize pets. Once nearly wiped out by pesticides, peregrine falcons (*Falco peregrinus*) have also moved into cities, including San Francisco, where they prey on other birds mostly, such as rock doves (*Columba livia*). They build nests in churches, window ledges of skyscrapers, and the towers of suspension bridges.

Human activities in natural areas stress nature animals. Even non-motorized activities can disrupt normal routines, affect feeding, and displace animals from their habitat. Some animals attempt to avoid humans, such as bobcat (*Lynx rufus*), grey fox (*Urocyon cinereoargenteus*), and mountain lion (*Puma concolor*). Others adapt to the presence and even thrive, such as coyote (*Canis latrans*) and raccoon (*Procyon lotor*). The results of these studies can help recreation park managers to limit the impact on wildlife while still allowing people the opportunity to experience wild areas. San Francisco is an excellent location to study these interactions.

In one study, Reilly et al. (2017) sought to determine how hiking, cycling, horse riding, and dog walking affected 10 species of mammals. They monitored camera traps at 241 locations in 87 protected areas to compare human and mammal activity in areas that were or were not used by humans and during day and night. When dogs were present, mountain lions and opossums were seen less frequently. Coyotes visited high-use areas at night, but not in the day. Small carnivores that hunt at night were not affected by daylight activities. They suggested prohibiting dogs and establishing buffers between areas used by humans and those by animals.

One challenge to restoring ground-nesting waterbird populations has been the removal of predators. Humans introduced some of those predators, including feral cats (*Felis domesticus*) and the Norway rat (*Rattus norvegicus*). Others are native to the Bay Area, such as the striped skunk (*Mephitis mephitis*). Meckstroth and Miles (2005) studied various strategies to lower the number and types of predator at several sites in the South Bay. They measured the number of nests and the hatching success in sites where predators were removed and sites where they were not removed. They found that nests in the removal sites had greater nest densities, but lower hatching success. The predator composition and abundance remained the same, except for feral cats. Striped skunks were the greatest predator and removal seemed to have no effect. They conclude that there are so many skunks that they easily repopulate removal sites. Their mixed results suggest that more study is needed to develop an effective plan to protect vulnerable species from predators.

INVASIVE SPECIES

A success in stopping an invasive species of salt marsh cordgrass illustrates excellent coordination between biology and government (Lubell et al., 2016). The cordgrass is a hybrid between introduced *Spartina alterniflora* and native *Spartina foliosa*. *Spartina* was introduced into the Bay by the Army Corps of Engineers in 1973. By 2005, it had increased to 805 acres in 12 areas. Its greatest amounts were found in the southeast portion of the Bay. To date, about 95% of the invasive species has been removed from the Bay.

In the 1960s, public awareness of the sorry state of San Francisco Bay began, and since then, the Bay has improved greatly (Briggs, 2016). The water is much more clear, attractive recreation areas have been established around the Bay, and habitats have improved for birds and mammals. However, fish and other marine populations have not returned to previous levels. For example, there are few large predators due to overfishing, and more than 200 invasive species have been introduced.

Native cordgrass (*Spartina foliosa*) is an important component of wetlands. For example, it traps sediment and provides nesting areas for the endangered California Ridgway's rail (*Rallus obsoletus obsoletus*). Nitrogen is limited in salt marshes, and adding nitrogen fertilizers has been a typical strategy for attempts to restore salt marshes. One such study (Murphy et al., 2018) tested two fertilizers for their ability to enhance cordgrass growth: kelp and sodium alginate. Although the alginate increased nitrogen fixation, there was no increase in cordgrass growth. The kelp lower nitrogen fixation but increased the nitrogen levels of *Salicornia pacifica* (a relative of pickleweed) where both plants are growing. The study also showed that phosphate was also a limiting factor for *S. foliosa* growth.

AIR QUALITY

In 2018, California suffered its worst wildfire season in some time. The Camp Fire in Butte County was the deadliest and most destructive fire in the state's history. The smoke from the fire blanketed much of Northern California for days, and at one point, San Francisco had the worst quality air in the world. In August 2020, an unusual thunder and lightning storm hit Northern California, resulting in 12,000 lightning strikes in the parched forests and grasslands in a week. A large number of wildfires again cause extremely poor air quality across the region.

Undoubtedly, wildfires caused bad air in the Bay Area periodically throughout history. Lightning strikes occurred, and volcanoes erupted to foul the air. However, air quality in the Bay Area must have remained fairly constant for most of its history, and for the most part, the air was pure. Later, Native Americans used fire for cooking, warmth, and to clear land for limited agriculture. The early Spanish colonist did the same. Yet air quality did not significantly deteriorate until after the Gold Rush and especially in the 20th century.

Controlling levels of dangerous gases is a challenge for most cities, including San Francisco and the Bay Area. Ozone, carbon monoxide, nitrogen oxides, particulates, and other chemicals are products of the modern world. Ozone formation is the result of meteorological conditions (e.g., temperature, water vapor in the air, and stagnant conditions) and emission (e.g., nitrogen oxides and volatile organic compounds).

Climate change adds another variable to the estimates of these gases in the future, and that new variable is likely to be significant in the near future. Models predict that, by 2050, the warming planet will result in significant changes to the hydrological cycle that will also affect the weather to lengthen stagnation events that leave bad air in place for extended times.

Air pollutants can be highly localized. For example, the Port of Oakland is in West Oakland and experiences high levels of ship and truck traffic as containers are brought in and out of the port. Thus, there residents are exposed to three times the levels of diesel exhaust as other residents in Oakland, and those pollutants are associated with poor health.

Auffhammer and Kellogg (2009) studied the effects of gasoline regulations on ozone levels in California. Those regulations add cost for customers and vary in their effectiveness in improving air quality. More specifically, they found that the federal regulations have no effect on emission of volatile organic compounds. However, the California regulations specifically require refiners to remove the chemicals that produce ozone, and thus, those regulations resulted in an improvement in air quality.

The fuel used by ocean-going ships contains high levels of sulfur that yield air pollutants, such as sulfur dioxide and particulates. In early 2009, California began to require ships to use lower sulfur fuels. Tao et al. (2013) looked at fine particle concentrations at four urban and two more remote sites in the Bay Area. Using vanadium as a marker for heavy fuel oil, they found that concentrations of that material were reduced at the urban sites and particularly at the West Oakland sites.

Climate change is likely to cause the air quality in the Bay Area to become worse in the near future (Steiner et al., 2013). Even now, excessive heat and drought have resulted in fires that made the air the worst in the world for a while in 2018. The continuing influx of people will also bring additional air pollution due to the continuing use of fossil fuels.

SEA LEVEL RISE

Sea level rise due to global warming is a serious threat to nearly all coastal areas. Estimates were that sea levels will rise 1.7 mm/year in the late 20th century to 3.1 mm/year in the early 21st century. Many low-lying areas will be at risk for inundation. Other aspects of climate change (e.g., storm intensity, rainfall, storm surges) will be exacerbated by climate change.

Scientists have been modeling these changes for some time. Different locations will be affected to greater or lesser degrees, and knowing what is likely to happen is critical for planning how best to adapt to those changes. Cloern et al. (2011) examined the Bay and particularly the Delta and Sacramento and San Joaquin Rivers. This region is at great risk for inundation as sea levels rise. The average sea level at the Golden Gate has risen 2.2 cm/decade since the 1930s, and extreme tides are 20 times more common since 1915. They predict that sea levels are likely to rise more rapidly than previously thought, and the water supply will diminish with wetter winters and drier summers. These factors will make is more difficult to maintain shorelines and native species in the future.

The Bay Area is using a "green" or ecosystem-based approach (Pinto et al., 2018). The large wetlands restoration projects add to the defense of the coast by providing additional areas for increased water levels to be absorbed even before they make contact with the seawalls, levees, and dikes that protect infrastructure and other items.

Local land subsidence may also make things worse, but maps displaying the risk of subsidence are lacking. To remedy this situation, Shirzaei and Bürgmann (2018) developed an accurate map of the risk of sea level rise in the San Francisco Bay. They used factors that indicate a larger rise than reported before. Their projections help identify coastal areas that will probably be flooded. They found some areas with subsidence rates of less than 2 mm/year and others that are greater than 10 mm/year. Those correlated with areas of artificial landfill or Holocene mud deposits. They predicted an area of 125–429 km^2 is at risk of flooding, which is somewhat higher than when only sea-level rise is considered. This more accurate mapping will help governmental agencies with planning for the future.

Subsidence can be caused by a number of factors, including not watering soil, compaction, pumping out groundwater, and removing crude oil. When it happens in levees, such as those in the Sacramento-San Joaquin Delta, it is a significant problem.

The use of metal-based coagulants to bind material together is one possible solution. To test this idea, one study (Stumpner et al., 2018) flooded wetlands with water containing coagulants that would accrete minerals. After 23 months of treatment, sediment samples were examined. Wetlands treated with poly-aluminum chloride worked the best, but iron sulfate was also good. Other characteristics of the accreted material were also encouraging. Overall, these coagulants seem to have efficacy in reversing subsidence and storing carbon. By using them, levee failure might be reduced.

HOPEFUL SIGNS

There are some very hopeful signs for the health of the Bay. For example, before 2008, harbor porpoises (*Phocoena phocoena*) were rare in the Bay. They were common until the 1940s when human activity in the Bay accelerated with the war effort. Now sightings are way up. In one study (Stern et al., 2017), more than 30 porpoises were seen every day. The average group size was 2.15, and the largest group was 16. About 10% of the groups were calves. The authors speculated that the reasons for their return might include less water and noise pollution, improved water quality, and more productive marine conditions for porpoises.

The Northern elephant seal (*Mirounga angustirostris*) is a magnificent animal. Adult males stand 14–16 feet and weigh up to 5500 pounds. Being true seals, they have no external ears and reduced limbs. Their name comes from the large proboscis of the males. The fur trade of previous centuries brought many species to near extinction. Elephant seals have few natural enemies, and so, they did not fear humans. This lack of fear made them an easy target for hunters who wanted their blubber for oil. As a result, they were nearly wiped out by the late 1800s. In 1922, a small group was found on Guadalupe Island off Baja California.

The first elephant seals arrived at Ano Nuevo just south of San Francisco in 1949. Under legal protection, the population has thrived, and now more than 150,000 seals return to 21 rookeries. Ano Nuevo has about 3000 each year. In recent years, another colony has formed further north at Drake's Beach at Point Reyes in Marin County. It now numbers about 600 individuals. The recovery of the Northern elephant seal population is a great success.

The California condor (*Gymnogyps californianus*) is a vulture and the largest land bird in North America. It is also an amazing story of recovery from near extinction. Their numbers plummeted in the 20th century. Poaching, lead poisoning from bullets that killed the carrion they feed on, and loss of habitat took a toll. In fact, it was completely extinct in the wild in 1987. The 27 living condors were captured in an effort to try to save the species at the San Diego Wild Animal Park and the Los Angeles Zoo. In 1991, they were reintroduced to the wild and have increased to 488 birds by 2018. A condor was spotted in San Mateo County in 2014.

The double-crested cormorant (*Phalacrocorax auritus*) is a common bird throughout North America, but in the late 19th century, their numbers in the Bay Area had plummeted to only one colony on the South Farallon Islands. By the 1970s, the population had increased to 20 colonies (Rauzoni et al., 2019). In the 1980s, they had established more than 1000 nests on the San Francisco-Oakland Bay Bridge and the Richmond-San Rafael Bridge. After the Bay Bridge was found to be seismically unsafe after the Loma Prieta earthquake in 1989, the eastern span of the bridge was replaced. The construction went on from 2003 to 2017, and during that time, the cormorant numbers decrease by 39%. A colony at Hog Island formed in 2001 and became the largest in the Bay Area. Nevertheless, the bridges are still major nesting areas for the birds. Although the bridge work disturbed them and their populations declined, the birds quickly moved to the new bridge once the old bridge was taken down and have become quite successful in exploiting these man-made structures for their purposes.

The mayors of those cities initiated a competition to see which city could cleanup the most trash from their coastal areas. The day involved 6400 volunteers who picked up more than 200,000 pounds of trash in just one morning. They also restored habitat and planted trees.

In 2016, voters in the Bay Area passed Measure AA, a regional parcel tax to help restore, enhance and protect wetlands. The measure passed with 70% approval. The US$500 million will be spent over the next 20 years to fund habitat restoration, flood protection, shoreline access, and recreational facilities.

Not all of the improvements to the Bay environment involve animals and plants. Removal of the Suisun Bay Reserve Fleet was a clear success story for improving the quality of the Bay. Established in 1946, the Fleet had at one point contained 340 ships that were berthed in Suisun Bay in case they might be needed in the future. However, the vessels were rarely activated and mostly were slowly decaying and polluting the Bay. In 2009, an agreement was reached to remove the ships, and the last one was towed away to be sold for scrap in 2017.

REFERENCES

Auffhammer M., Kellogg R. (2009) Clearing the air? The effects of gasoline content regulation on air quality. *American Economic Review* 101: 2687–2722.

Barnard P.L., Foxgrover A.C., Elias D.P.L., Erikson L.H., Hein J.R., McGann M., Mizell K., Rosenbauer R.J., Swarzenski P.W., Takesue R.K., Wong F.L., Woodrow D.L. (2013) Integration of bed characteristics, geochemical tracers, current measurements, and numerical modeling for assessing the provenance of beach sand in the San Francisco Bay Coastal System. *Marine Geology* 336: 120–145.

Barnard P.L., Schoellhamer D.H., Jaffe B.E., McKee L.J. (2013) Sediment transport in the San Francisco Bay Coastal System: An overview. *Marine Geology* 345: 3–17.

Bever A.J., MacWilliams M.L., Wu F., Andes L., Conner C.S. (2014) Numerical modeling of sediment dispersal following dredge material placements to examine possible augmentation of the sediment supply to marshes and tidal flats, San Francisco Bay, USA. *33rd PIANC World Congress*. World Association for Waterborne Transport Infrastructure, Alexandria, VA.

Bible J.M., Sanford E. (2016) Local adaptation in an estuarine foundation species: Implications for restoration. *Biological Conservation* 193: 95–102.

Brew D.S., Williams P.B. (2010) Predicting the impact of large-scale tidal wetland restoration on morphodynamics and habitat evolution in South San Francisco Bay, California. *Journal of Coastal Research* 26: 912–924.

Briggs J.C. (2016) San Francisco Bay: Restoration progress. *Regional Studies in Marine Science* 3: 101–106.

Capello M., Cutroneo L., Castellano M., Orsi M., Pieracci A., Bertolotto R.M., Povero P., Tucci S. (2010) Physical and sedimentological characterisation of dredged sediments. *Chemistry and Ecology* 26:sup1, 359–369.

Chin J.L., Ota A. (2001) Disposal of Dredges Material and Other Waste on the Continental Shelf and Slope. In: *Beyond the Golden Gate: Oceanography, Geology, Biology and Environmental Issues in the Gulf of the Farallones*, edited by Herman A. Karl, 62–65. Denver, CO: United States Geological Survey.

Cho Y.-M., Werner D., Choi Y.J., Luthy R.G. (2012) Long-term monitoring and modeling of the mass transfer of polychlorinated biphenyls in sediment following pilot-scale *in-situ* amendment with activated carbon. *Journal of Contaminant Hydrology* 129–130: 25–37.

Cloern J.E., Knowles N., Brown L.R., Cayan D., Dettinger M.D., Morgan T.L., Schoellhamer D.H., Stacey M.T., van der Wegen M., Wagner R.W., Jassby A.D.

(2011) Projected evolution of California's San Francisco Bay-Delta-River system in a century of climate change. *PLoS One* 6(9): e24465.

Davis J.A., Looker R.E., Yee D., Marvin-DiPasqualec M., Grenier J.L., Austin C.M., McKee L.J., Greenfield B.K., Brodberg R., Blum J.D. (2012) Reducing methylmercury accumulation in the food webs of San Francisco Bay and its local watersheds. *Environmental Research* 119: 3–26.

De La Cruz S.E.W., Smith L.M., Moskal S.M., Strong C., Krause J., Wang Y., Takekawa J.Y. (2018) Trends and Habitat Associations of Waterbirds Using the South Bay Salt Pond Restoration Project, San Francisco Bay, California. *U.S. Geological Survey*. Open-File Report 2018-1040, 136 p. https://doi.org/10.3133/ofr20181040.

Elias E.P.L., Hansen J.E. (2013) Understanding processes controlling sediment transports at the mouth of a highly energetic Inlet system (San Francisco Bay, CA). *Marine Geology* 345: 207–220.

Foxgrover A.C., Marvin-DiPasquale M., Jaffea B.E., Fregoso T.A. (2019) Slough evolution and legacy mercury remobilization induced by wetland restoration in South San Francisco Bay. *Estuarine, Coastal and Shelf Science* 220: 1–12.

Fraser M., Short J., Kendrick G., McLean D., Keesing J., Byrne M., Caley M.J., Clarke D., Davis A., Erftemeijer P., Field S., Gustin-Craig S., Huisman J., Keough M., Lavery P., Masini R., McMahon K., Mengersen K., Rasheed M., Statton J., Stoddart J., Wu P. (2017) Effects of dredging on critical ecological processes for marine invertebrates, seagrasses and macroalgae, and the potential for management with environmental windows using Western Australia as a case study. *Ecological Indicators* 78: 229–242.

Ganju N.K. (2019) Marshes are the new beaches: Integrating sediment transport into restoration planning. *Estuaries and Coasts* 42: 917–926.

Gies E. (2018) Fortresses of mud: How to protect the San Francisco Bay Area from rising seas. *Nature* 562: 178–180.

Greb S.F., DiMichele WA, and Gastaldo R.A. (2006), Evolution and importance of wetlands in earth history. *Geological Society of America (Special Papers)* 399: 1–40.

Hartman C.A., Ackerman J.T., Herzog M.P., Strong C. Trachtenbarg D. (2019) Social attraction used to establish Caspian tern nesting colonies in San Francisco Bay. *Global Ecology and Conservation* 20: e00757.

Love A.H., Esser B.K., Hunt J.R. (2003) Reconstructing contaminant deposition in a San Francisco Bay marine, California. *Journal of Environmental Engineering* 129: 659–666.

Lubell M., Jasny L., Hastings A. (2016) Network governance for invasive species management. *Conservation Letters* 10: 699–707. https://doi.org/10.1111/conl.12311.

Matella M.K., Merenlender A.M. (2014) Scenarios for restoring floodpain ecology given changes to river flows under climate change: Case from the San Joaquin River, California. *River Research and Applications* 31: 280–290.

Meckstroth A.M., Miles A.K. (2005) Predator removal and nesting waterbird success at San Francisco Bay, California. *Waterbirds: The International Journal of Waterbird Biology* 28: 250–255.

Milligan B., Holmes R. (2017) Sediment is critical infrastructure for the future of California's Bay-Delta. *Shore & Beach*, 85(2).

Morris J.D. (2020) 50,000 gallons of sewage spill into Oakland-Alameda waters after power failure. *San Francisco Chronicle*. https://www.sfchronicle.com/bayarea/article/50-000-gallons-of-sewage-spill-into-15486932.php, accessed December 7, 2020.

Murphy J.L., Boyer K.E., Carpenter E.J. (2018) Restoration of cordgrass salt marshes: Limited effects of organic matter additions on nitrogen fixation. *Wetlands* 38: 361–371.

Perelo L.W. (2010) In situ and bioremediation of organic pollutants in aquatic sediments. *Journal of Hazardous Materials* 177: 81–89.

Pinto P.J., Kondolf G.M., Wong P.L.R. (2018) Adapting to sea level rise: Emerging governance issues in the San Francisco Bay Region. *Environmental Science & Policy* 90: 28–37.

Rauzoni M.J., Elliott M.L., Capitolo P.J., Tarjan L.M., McChesney G.J., Kelly J.P., Carter H.R. (2019) Changes in abundance and distribution of nesting double-crested cormorants *Phalacrocorax auratus* in the San Francisco Bay Area, 1975-2017. *Marine Ornithology* 47: 127–138.

Reilly M.L., Tobler M.W., Sonderegger D.L., Beier P. (2017) Spatial and temporal response of wildlife to recreational activities in the San Francisco Bay ecoregion. *Biological Conservation* 207: 117–126.

Sacks B.N., Ernest H.B., Boydston E.E. (2006) San Francisco's Golden Gate: a bridge between historically distinct coyote (*Canis latrans*) populations? *Western North American Naturalist* 66: 2, Article 16.

San Francisco Estuary Partnership, (2015). State of the Estuary Report. https://www.sfestuary.org/wp-content/uploads/2015/10/SOTER_2.pdf, Last viewed: December 7, 2020.

Shellenbarger G.C., Wright S.A., Schoellhamer D.H. (2013) A sediment budget for the southern reach in San Francisco Bay, CA: Implications for habitat restoration. *Marine Geology* 345: 281–293.

Shirzaei M., Bürgmann R. (2018) Global climate change and local land subsidence exacerbate inundation risk to the San Francisco Bay Area. *Science Advances* 4: eaap9234.

Steiner A.L., Tonse S., Cohen R.C., Goldstein A.H., Harley R.A. (2013) Influence of future climate and emissions on regional air quality in California. *Journal of Geophysical Research* 111: D18303.

Stern S.J., Keener W., Szczepaniak ID., Webber M.A. (2017) Return of harbor porpoises (*Phocoena phocoena*) to San Francisco Bay. *Aquatic Mammals* 43: 691–702.

Stralberg D.M., Brennan M., Callaway J.C., Wood J.K., Schile L.M., Jongsomjit D., Kelly M., Parker V.T., Crooks S. (2011) Evaluating tidal marsh sustainability in the face of sea-level rise: a hybrid modeling approach applied to San Francisco Bay. *PLoS One* 6: e27388.

Stumpner E.B., Kraus T.E.C., Liang Y.L., Bachand S.M., Horwath W.R., Bachand P.A.M. (2018) Sediment accretion and carbon storage in constructed wetlands receiving water treated with metal-based coagulants. *Ecological Engineering* 111: 176–185.

Tao L., Fairley D., Kleeman M.J., Harley R.A. (2013) Effects of switching to lower sulfur marine fuel oil on air quality in the San Francisco Bay Area. *Environmental Sciences & Technology* 47: 10171–10178.

Wilber D., Clarke D. (2010) Dredging Activities and the Potential Impacts of Sediment Resuspension and Sedimentation On Oyster Reefs. *Proceedings of the Western Dredging Association Technical Conference, June 6-9, 2010*, San Juan, Puerto Rico, USA, 61–69.

Yee, D. Wong, A. 2019. Evaluation of PCB Concentrations, Masses, and Movement from Dredged Areas in San Francisco Bay. *SFEI Contribution #938*. San Francisco Estuary Institute, Richmond, CA.

10 Future of the Bay

On April 17, 1906, San Francisco was a thriving city. It had grown up from its sudden beginning in the Gold Rush days. With about 400,000 people, it was the largest city on the West Coast and the 7th largest in the United States. In addition, it was the most important port and financial center in the West and had many of the features of a great city: the largest hotel and civic building in the West. The US Mint at San Francisco held about one-third of the nation's gold supply. The great tenor Enrico Caruso had performed Bizet's *Carmen* at the Mission Opera House that evening.

On the very next day at 5:12 am, when most of its residents were still asleep, the city was struck by a strong earthquake, later estimated to be magnitude 7.8. One of those was Caruso, who was sleeping in The Palace Hotel. The shaking and fire that followed killed about 3000 people, destroyed 80% of the city, and left half of the city's residents homeless. That natural disaster showed that even the solid ground that we take for granted can be gone in a moment. And it also demonstrates that one of the major forces that built the Bay is still active and, in fact, presages its ultimate fate.

Of course, today San Francisco has long been rebuilt and has become one of the great cities of the world. The hills and water that surround the city are a wonder of nature. Beauty is everywhere. From the Marin headlands to Treasure Island to the Berkeley hills. However, as we have seen in the earlier chapters, the Bay was not always here, and it will not always be here in the future. The same forces that built the Bay are still at work, and eventually, they will destroy it. It will take millions of years. Although humans saw the beginning of the Bay, it is far from clear that we will see the end of it.

We are periodically reminded of some of the forces that built the Bay. Earthquakes occur unexpectedly, and jolt us back into awareness that we could experience "the big one" at any time. In fact, as we move into the future, the Bay Area will have many big ones. Since they happen only every 100–200 years, we can be fooled into thinking they are rare. They are rare for humans. However, geological forces are on a different, longer timeline, and the future is open-ended. Time moves on forever. In a million years, we could have 5000 big ones. And it took 10 million years to build the Bay Area.

Other forces are at play too. Perhaps they are less dramatic, but they will also affect the Bay Area. Some are gone. The volcanoes that were active several million years ago are now extinct. The energy powering them was the Juan de Fuca plate, which has slid northwest and now powers the Cascade volcanoes in Oregon and Washington. However, other forces, such as human activities, climate change, sea level rise, flooding, and others continue to act. In this chapter, we will look at them in some detail.

PEOPLE AND MORE PEOPLE

Humans have lived in the San Francisco Bay Area for about 10,000 years or so. According to the US Census Bureau, about 7.5 million people live in the San Francisco Bay Area, and that number grows by about 100,000 each year. By 2040, the population will be 9.3 million! They will have housing, food, water, sanitation, transportation, recreation, and other needs that will put even more pressure on the Bay.

For the last 150 years, we humans have had an enormous effect on the Bay Area. The early gold mining practices transformed mountains into silt that was later washed into the Bay. We have built cities, roads, and bridges. We filled a significant portion of the Bay itself and built on it. Wastes and pollutants have been poured into the Bay. Pesticides, mercury and other metals, toxic substances, and invasive species are at unacceptable levels in many parts of the Bay. Bacteria due to sewage spills and deteriorating sewage infrastructure pollute the beaches. In the last 150 years, people have destroyed 90% of the wetlands and 40% of the aquatic ecosystem. In recent years, efforts have been made to protect the Bay and to preserve and even expand the number of wetlands and tidal marshes.

Humans have also introduced new "alien" species to the Bay that are a threat to the native species. A study by Molnar et al. (2008) showed 85 invasive species in the Bay. Many of these come in the ballast of ships coming to the Bay Area, and they include Asian clams, Chinese mitten crabs, Amur River clams, New Zealand carnivorous sea slugs, and Black Sea jellyfish. Commercial shipping is increasing, so the effects of these invasive species are also increasing.

Humans will continue to factor into the future of the Bay. Major construction will be needed to supply the housing and other infrastructure to support the growing population, and tensions are likely to increase between the needed expansion of the human footprint and the efforts to preserve the natural environment of the Bay. Already there is extreme pressure on many wild species, and it is hard to envision how native plants and animals will survive the increasing density of humans.

EROSION AND SILTING

Erosion is a natural process. Wind, water, and changing temperatures slowly erode the rocks that make up mountains. The resulting debris is then down California's rivers for millennia. Eventually, that silt reached the Bay.

Gold mining in the late 19th century changed all of that. Hydraulic mining was particularly destructive of the natural environment. High-power jets of water were used to eat away entire mountains to get at the gold. Trillions of cubic feet of gravel, sand, and mud were washed down from the mountains into the delta and Bay. As the rivers reached the valley, the water flow slowed, and the silt choked rivers and streams and buried valleys. Eventually, the massive amount of silt reached the Bay and reduced its depth and clarity. The silt was a mixed blessing. It hindered navigation on the rivers and the Bay.

But for the last 150 years, it has provided a home for many plants and animals in the Bay.

However, in the late 20th and early 21st centuries, much of the sediment has been washed out to sea. More recently, dams and the installation of riprap on the banks of rivers, particularly the lower Sacramento River, have greatly reduced the amount of silt arriving in the Bay. Less sediment in the water allows more sunlight to penetrate to deeper depths. The increased sunlight encourages the growth of phytoplankton, which can result in a massive die-off of other plants. When those dead plants sink and begin decomposing, the oxygen levels in the water are depleted to the detriment of fish and other animals. This will have significant implications for the marine life in the Bay. Certain species, such as the endangered Delta smelt, need water with some sediment. They and other fish use the sediment as a sort of camouflage.

Other human actions added to the loss of sediment. The naturally occurring levees were built up to make them permanent. Many of the islands in the broad delta were made of peat. Intensive farming quite literally used up the peat-soil so that the islands have sunk 5–6 s meters below sea level. Sea level rise and earthquakes are significant threats to the levees.

Just outside the Golden Gate is a large sand bar that was deposited over centuries. The sand bar helps to protect the Bay from the action of the Pacific by absorbing much of the wave energy. Sand is deposited there and taken away by the action of seawater. In the future, less new sediment will arrive at the sand bar, and sea levels will rise. The combination of these two actions will cause the sand bar and its protection to be diminished. Consequently, the Bay will be exposed to more severe wave action than in the past.

Ruckert et al. (2017) examined the risk of flooding in the San Francisco Bay due to sea level rise. Most analyses look at the expected, best, or large quantile (i.e., 90%) estimate because the uncertainty involved in these estimates can be significant. Ruckert et al. delved into the uncertainty to account for that additional risk. They found that, by 2100, the area that is likely to be inundated by sea level rise is twice the current estimates.

CLIMATE CHANGE AND TOO MUCH WATER

Climate change has occurred throughout the history of the Earth. However, since the beginning of the Industrial Age, human activities have had a major influence on global warming. The burning of fossil fuels has injected large amounts of carbon dioxide, and other human activities have released more methane in the atmosphere. Both are greenhouse gases that trap heat from the sun on the Earth. As the average temperatures rise, the polar caps and large glaciers begin to melt and run off into the oceans. Without the reflection of the ice and snow, the surface albedo decreases, more sunlight is trapped, and the whole process accelerates.

Global warming is a threat to many organisms, and its effects are already being seen in the animals around the Bay. Unfortunately, greenhouse gas emissions continue to rise and so does the average global temperature. Every living thing is affected. Amphibians are perhaps the more endangered. Nearly a third of all amphibians are at risk right now (Wake and Vredenburg, 2008). Climate change is part of the threat, but other causes include loss of habitat, pollution, exotic species, and disease. Amphibians may be the most threatened, but they are not something that we see every day. Birds are. So, the loss of birds is more visible to us. In fact, they are quite threatened. Since 1970, we have lost 3 billion birds or 29% of all birds (Rosenberg et al., 2019). Fish are also threatened by dams and other infrastructure projects, pollution, and invasive species. The American Fisheries Society estimates that 39% of all North American freshwater and anadromous fish are in trouble (Jelks et al., 2008). Invertebrates are the most common organisms on the Earth and the most diverse. They account for 97% of all species. The loss of bees is a great example. They are critical for the pollination of a large majority of human food crops (Klein et al., 2018). Mammals, and particularly primates, are under great stress. Animals are not the only organisms threatened by climate change. Plants are also under pressure. They cannot move to find more suitable environments (Tilman et al., 1994).

We are already witnessing changes due to climate change, and they represent a major threat to all of us. By the year 2200, scientists estimate that our atmosphere will contain higher levels of carbon dioxide than any time over the last 650,000 years. The warming will accelerate the melting of the polar ice caps and cause the levels of the oceans to rise even faster. Thus, the water arriving in the Bay will come from more sources than just rain or upstream run-off. However, the rise in any specific area is not easy to predict.

The mean sea level in San Francisco Bay has risen 22 cm over the last century. Knowles (2010) attempted to determine the areas of the Bay that would be inundated whether they have levees or not. Even assuming no sea level rise, some low-lying areas are at risk of flooding under extreme conditions. Rising sea levels simply add to the area at risk (Fig. 10.1). Knowles looked at likely flooding between two extremes: no sea level rise and 150 cm of rise. Estimates of the actual rise by 2100 vary between 50 and 150 cm. Those numbers agree well with a US Geologic Survey (USGS) estimate that sea level in the Bay Area will rise by 1.24 m by 2100 (Takekawa et al., 2013). Some developed areas are at risk but, for the most part, are protected by levees. With rising sea levels, many of those developed areas might flood as levees fail or are overtopped. Those areas include both main airports for San Francisco and Oakland and many other areas of municipal or industrial development. Most analyses look at the expected, best, or large quantile (i.e., 90%) estimate because the uncertainty involved in these estimates can be significant. Some estimates (Ruckert et al., 2017) examined the risk of flooding is much greater. They found that, by 2100, the area that is likely to be inundated by sea level rise is twice the current estimates.

As the ice melts and runs off to the oceans, sea levels rise, and coastal communities are threatened with inundation and storm and tidal surge damages. The loss of highly populated and developed coastal areas will be very disruptive

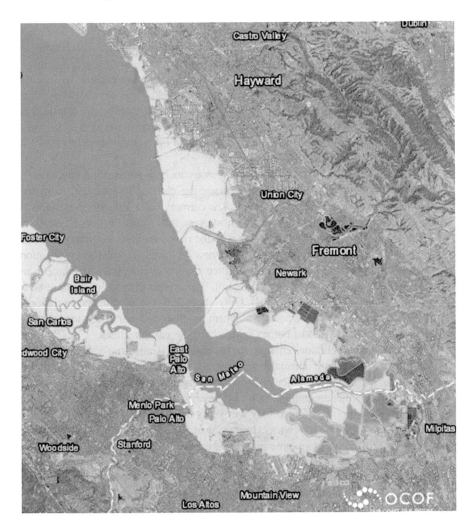

FIGURE 10.1   Sea Level Rise. Climate change is causing the polar ice caps and permanent glaciers to melt to a greater extent than usual. As more ice melts, sea levels rise, and low-lying areas in the Bay are at risk of flooding.

to those societies. Local and state governments will have to decide how to adapt to the rising waters. Low-lying area at the coasts will be vulnerable, and the San Francisco Bay Area will feel the rise acutely. Much of the Bay Area is only a few feet above sea level now. In fact, large swaths of territory now built on was recovered from the Bay by filling the shallow areas. It will not take much ocean rise for the Bay to reclaim those areas. New barriers can be built. Communities and infrastructure might need to be relocated. For example, the three major airports in the Bay Area are all under threat. The cost will be staggering. Other consequences may occur, including damage to sanitation systems. Governmental agencies will have to rethink development in low-lying areas, and insurance companies will need to reevaluate risk for development. Ground water extraction will need to be monitored to insure against subsidence. The areas that will feel the effects of sea level rise the earliest will be the marsh areas. They will be inundated and lost. Takekawa et al. (2013) estimate that 95% of the marsh area will be flooded by that rise and lose its marsh plant communities. About 96% of the marshes in San Francisco Bay will be overwhelmed by the rise in sea level and will become open water mudflats by 2100 (Takekawa et al., 2013).

Tidal wetlands are important for many reasons, including water filtration, slowing floods, protecting infrastructure, and sequestering carbon. To maintain them will require an enormous amount of material. The current rate of sediment deposition on the wetlands is not nearly enough for most areas that would be flooded. In some places, wetlands might be able to move with the rising sea level, but many areas are bounded by development that would limit new wetlands. To maintain navigation, the Bay must be continually dredged. Three million cubic yards of sediment are dredged each year and must be disposed of. The dredge material could be used to help marshlands (Bever et al., 2014). Only about 20% can be dumped in the open ocean. So about 40% of it is used to augment mudflats and marshes around the Bay and to create new wetlands, but no solution will be easy or cheap.

Even ignoring climate change for a moment, California's Mediterranean climate makes the state unusually sensitive to drought and flooding. The amount of rain can be greater or lesser in any given year. Atmospheric rivers are known to bring exceptional amounts of rain to California. Some studies have shown that the state will experience more very wet years in the future (Swain et al., 2018), such as the year 1862 that was extremely wet.

Extreme differences in precipitation are often associated with atmospheric rivers. Gershunov et al. (2019) examined 16 global climate models to determine how they predict atmospheric rivers. The five most accurate models predict great variability in precipitation in the Western US and particularly in California. They did find that any increase in rainfall is due almost entirely to atmospheric rivers. However, O'Gorman (2015) points out that the effects of precipitation extremes are not simple. Some factors, such as convection effects and the duration and type of precipitation, are not well understood. Climate change contributes another confounding factor to the mix.

FUTURE DROUGHTS

The Bay Area has experienced droughts for much of its existence, and the last few years have been the most severe droughts on record (Griffin and Anchukaitis, 2014). Is this simply a one-off anomaly or does it predict even more severe droughts in the future? Will climate change and the increase in global warming lengthen and deepen future droughts? Some studies have indicated that human activities and global warming have already enhanced the present droughts (Diffenbaugh et al., 2015).

Cook et al. (2015) sought to answer these questions. They agree with previous studies that climate change has exacerbated current droughts to make them worse than the Medieval-era droughts and the worst in history. Then they attempt to predict what the future will look like. Unfortunately for the Bay Area, they predict that the second half of the 21st century will be much drier than the 20th century or previous times. They believe that it will be even drier than the Medieval Climate Anomaly (1100–1300 CE).

Water has traditionally been the subjects of significant disputes. There is barely enough to support the current competing demands of a growing population, the enormously successful agricultural industry, and the needs of the environment to sustain plants and animals. The infrastructure that exists struggles to serve all of these. Was California's water infrastructure built under the assumption of water would be more plentiful than it actually will be and what might the consequences of that scenario be? Harou et al. (2010) built a model of a 72-year drought with half the normal precipitation and used it to examine the effects of such a drought on California in 2020 and the state's response to it. They conclude that the state could adapt its economy to the drought. However, the agricultural concerns and environmental needs would be severely disrupted. In fact, the loss of water might be the end of some ecosystems.

The recent droughts featured a significant loss of trees. That is not surprising, but the actual reason is not known. Goulden and Bales (2019) sought to determine that reason. They studied the recent drought years of 2012–2015. The moisture was depleted to a depth of 5–15 m. Dense vegetation and transpiration added to the effect. Climate change is expected to make die-offs even worse in future years.

WILDFIRES

For the last few years, California and the West Coast have experienced a dramatic increase in the number of wildfires. Those fires killed over 150 people, destroyed 30,000 buildings, and burned 1.2 million hectares. The economic losses were staggering.

Many factors are involved in the increase in fires. The wild areas are beautiful, and lots of people want to live there. Unfortunately, those areas are at greater risk for fire. Forest management practices have limited the number of smaller fires and allowed the burden of flammable material to accumulate over the decades.

However, the biggest reason for more and more severe fires is climate change (Goss et al., 2020). The climate in California and the Bay Area has changed over the last few decades. Average temperatures have increased. Droughts are longer, and plants are stressed for lack of water. The snowpacks in the Sierras, which are so important in the Bay Area, are less. All of these factors have combined to make the Bay Area's fire season much longer than it was just 20 years ago. Smoke covers nearly the entire state, and the air quality is very bad.

Importantly, the great majority of fires are caused by humans. Climate change dramatically increases the risk for fires, but the spark that ignites them comes

FIGURE 10.2   Darkness at Noon. In August 2020, the smoke from multiple wildfires blanketed nearly all of California. In this photograph, the smoke turned the sky in Sacramento orange, and cars had to drive with lights on at mid-day.

mostly from humans. About 88% of the fires and 92% of the burned area come from incidents started by people. There are occasional exceptions. For example, on one weekend in August 2020, more than 11,000 lightning strikes set hundreds of wildfires (Fig. 10.2). However, this is the so-called example that proves the rule. Lightning storms are very rare in the Bay Area. Most fires are caused by people.

Climate change in general, and wildfires in particular, are significant threats to living organisms throughout the Bay Area and California. Many species have little tolerance for the higher temperatures and longer droughts. Fires are the ultimate threat. The smoke and flames kill outright, but the loss of habitat after the fire is also stressful. As one example, Wan et al. (2019) studied the effects of climate change and wildfire on spotted owl populations. The Mexican spotted owl is expected to have a 13-fold increase in the amount of its habitat lost to fires by 2080.

LANDSLIDES

The movement of earth depends on the amount of water in the soil. Landslides have been modeled carefully by mechanical models, but their behavior in the field is less well-understood. Handwerger et al. (2019) undertook to study landslides and water in the field. Using radar interferometry and a simple hydrological model, they found that the risk of landslide was lower during the drought and increased greatly with the rains. The increase of precipitation extremes with climate change will greatly increase the risk of landslides.

Climate change will also encourage rockslides. Those that have been moving slowly can suddenly fail with serious implications for people and property. Increasing pore pressure by fluids in the soil increase creep rates and can reach a stochastic point at which the system fails (Agliardi et al., 2020).

FUTURE OF THE DELTA

At the end of the last Ice Age, the Sacramento River Delta formed when sediments from the San Joaquin and Sacramento Rivers behind the Carquinez Strait. For the last 10,000 years, the Delta existed as a large freshwater marsh with lots of channels and sloughs around islands of peat and tule. Over that time, the tules and cattails that formed the bulk of the vegetation in the Delta added only a few inches of new soil each year. Since the mid-19th century, much of the Delta has been used for agriculture.

The land is very rich for farming. However, the farmers built dikes and levees and drained parts of the Delta. Those actions expose the soil to oxygen, which encourages the aerobic growth of bacteria and other microorganisms that eat the rich carbon-containing material in the soil. The loss of that organic material makes the soil less dense, and it tends to shrink and become compacted. As a result, the levees lose height and density. They become weaker and more vulnerable to damage or breaching in storms. They must be built up higher and strengthened or the fields, homes, and other structures on the island will be threatened by storms.

The climate crisis is increasing the risk. Sea levels are rising. Some estimates are as high a meter of rise in this century and more after that. Right now, about 1150 square miles of the Delta are under sea level. Many of the levees will not be able to protect against the sea level rise and storms. In fact, without significant improvements to the levee system, the Delta might be transformed into a huge lake of 1200 square miles.

The process has already begun. Franks Tract was owned by Dr. N.K. Foster and F.C. Franks, and later used to grow potatoes, beans, asparagus, sugar beets, onions, seed crops, small grains, and corn. It was also a very productive peat mine, which did not help its subsidence. In 1937, the levee failed, and the land was flooded. It was reclaimed shortly thereafter, but flooded again in 1938. It was never reclaimed and is now a state park. Mildred Island flooded in 1983. The water in parts of the former island is now 15 feet deep. In 1998, a levee fails and Liberty Island is inundated. In 2004, Jones Tract levee fails and is repaired at great expense.

The islands can be reclaimed in cases. Suddeth et al. (2010) concluded that repairing the levees around 34 islands in the Delta that were examined are economically not feasible and that the repairs would likely not improve the survivability of the islands anyway. Others have become fairly productive fishing and recreational areas. However, this is not always successful.

Moreno-Mateos et al. (2012) showed that, even when the islands were carefully restored, they were much less productive. They conducted a meta-analysis of 621 wetland sites around the world. They found that the biological structure and biogeochemical functioning of the sites were, on average, 26% and 23% lower, respectively, than reference sites. Restoring the islands is a bit like trying to put Humpty Dumpty together again.

A NEW ICE AGE?

The Earth alternates between periods of warmer and cooler conditions. It has had several ice ages in the past and will have more in the future. The cooler times occur about every 200 million years and last for millions or tens of millions of years. We call them ice ages, and they result in increased glaciation and lowered sea levels. Even within an ice age, there are warmer and cooler periods. Interestingly, although it might not be immediately apparent, we are currently in an Ice Age. Its peak was 20,000 years ago, but fortunately for us, we are in one of the warmer periods of this ice age.

What causes the ice ages is not completely clear. There are a number of ideas. Periodically, the Earth's orbit changes from more or less a circle to an ellipse, and the Earth tilts a little more. The great Serbian scientist Milutin Milanković first suggested that changes in the Earth's orbit caused the Ice Ages. These variations may change the amount of sunlight that strikes the Earth. Although these are relatively small "wobbles" in the orbit, they are enough to cause drastic changes in the Earth's temperature and initiate an Ice Age. In fact, their regularity can be used to predict the timing of the next Ice Age (Hays et al., 1976). Using that model, scientists estimate that the next Ice Age will begin in about 1500 years. Plate tectonics may also be involved. Changes in the opens may allow or preclude ocean circulation. Changes in the amount of carbon dioxide in the atmosphere are another strong candidate. Plate collisions that result in new mountain chains might also disrupt the circulation of the atmosphere. However, the situation is now complicated by the increase of greenhouse gases into the atmosphere by human activity since agriculture began and particularly in the last 200 years with the Industrial Age (Maslin, 2016; 2020). Some have wondered if the increase in global temperatures might offset the effects of the changes to the Earth's orbit and postpone the next Ice Age.

The cooling of the Earth results in an Ice Age, and ice ages generally last about 100,000 years with a warmer period of 10,000 years in between. We have been in one of the warmer periods for some time. Some scientists believe we are moving to the end of that period and that a new Ice Age will begin. That will result in larger ice caps and larger mountain glaciers. More water will be frozen, and less will be in the oceans. Sea levels will drop. During the last Ice Age, all of Canada and New England and much of New York were covered by ice, as well as far south as Missouri.

Another type of ice age might influence the Bay on a much shorter timeline. This "little ice age" would be the result of a grand solar minimum, in which the total solar irradiance is reduced by 0.25%. Some scientists predict that such an event will occur between 2020 and 2070. One of these, called the Maunder Minimum, cooled the Earth in 1650–1700. Other examples are known throughout the history. Current scientific thought seems to indicate that this event will slow but not fully correct the influence of human activities on global warming.

PLATE TECTONICS AND EARTHQUAKES

By far, the most powerful of the forces acting on the Bay Area is plate tectonics. Just like the Bay itself, the continents that exist now have not been that way forever, and they will not be that way far into the future. The continents will change radically, and one of the casualties of all of that change will be the Bay Area. Those of us who live in the Bay Area are periodically reminded that we live in a geologically active region by earthquakes. Over the next millions of years, the cumulative result of all of those earthquakes, large and small, will be that part of California will continue to slowly move northward. In about 100 million years, "San Francisco" will be just off the west coast of Canada and nearing Alaska.

The Earth and all of its components are in constant motion and ever changing. The energy for that motion is supplied by the Earth's superhot core, which is composed mostly of iron. The inner core is a solid mass at 7600–13,000 °F. The outer core is molten iron at 5800–9400 °F. The movement of the plates is powered by spreading in the middle of the oceans as new material upwells from the mantle and is later subducted where plates meet. Along those edges, earthquakes and volcanoes are common. Ultimately, the energy for the movements comes from the Earth's cooling, but still, fantastically hot mantle and core.

There is a rhythm to plate tectonics that causes the continents to alternate between a huge single supercontinent to multiple smaller continents. A supercontinent lasts about 100 million years before it breaks up into several continents that drift apart. After a few hundred million years, the continents merge again to form another supercontinent.

Scientists believe that three supercontinents have likely existed over the past 2 billion years. Other supercontinents might have preceded those three, but the evidence is weaker as we try to see more deeply into the past. The oldest supercontinent known so far, Nuna, existed about 1.8 billion years ago, Rodinia was about 1 billion years ago, and Pangaea was about 300 million years ago. In the times between those three, varying numbers of smaller continents, such as we have today, existed. Some of the evidence for this movement comes from rocks containing iron. Iron can be magnetized, and when the rock was molten, the magnetized iron oriented itself along the magnetic lines of the Earth. Once the rock solidified, its history was locked in stone and can be compared to the magnetic orientation of other rocks.

Scientists seem to be convinced that the next supercontinent will involve the America and Asia. Three computer models give similar answers, but one

(Mitchell et al., 2012) suggests that the Caribbean Sea and the Arctic Ocean will disappear, and South America will wind up alongside the eastern coast of North America. The relatively dense oceanic crust flows under the continental crust so that the huge land masses move over the surface of the earth much like water runs downhill.

One of the factors involved is delamination (Putirka and Busby, 2011). The crust and mantle experience a lot of movement. Part is due to plate tectonics, but buoyant density is also involved. The lithosphere includes the crust and top part of the mantle. The Asthenosphere is the upper layer of the mantle, just under the lithosphere. It is a semi-solid but flows. During subductions, relatively cold continental mantle lithosphere is more dense than the asthenosphere below it. This inversion causes the lithosphere to thin and allows the hotter asthenosphere to move upwards. Delamination occurred as the Farallon plate was subducted under the North American plate. In that example, the uplift resulted in the mountain building of the Sierra Nevada and the volcanic activity in eastern California. Those geologic activities still cause uplift and depressions today, and in that way, they can relieve some of the pent-up stress on the faults. Delamination also occurs at rift zones.

Most of us—at least those of us who live on the West Coast—experience plate tectonics through its most violent manifestation, earthquakes. In some ways, they seem uncommon. In the Bay Area, earthquakes are felt every year. But they are really very common. Becker and Geschwind (2016) posted a dramatic demonstration of the major earthquakes around the world from 2001 to 2015. Their documentation clearly shows the "ring of fire" that circles the Pacific Ocean and the other major faults and rifts. The video provides a far better appreciation of the movement along those faults than any words. Part of the Pacific "ring of fire" includes the US West Coast. The San Andreas fault system is a continental transform fault of about 800 miles that extends from Southern California near the Salton Sea and runs roughly northwest to nearly Eureka in Northern California. At San Francisco, the fault runs offshore and then back on shore just north of the city. The fault is the boundary between the North American and the Pacific plates. Both are moving. The Pacific plate is moving northwest and continues to grind against the North American plate, which is moving southwest and pushing on the Pacific plate. The movement is sometimes smooth and sometimes sudden, resulting in an earthquake. Over millions of years, the most western part of California has drifted north from the Los Angeles area to form part of the peninsula and part of the city of San Francisco. The Pinnacles National Park just east of Monterey, California, is a great example of this movement. The Pinnacles are tall rock formations that are the leftovers of a large volcano. The softer rock around has eroded away, leaving the columns of very hard volcanic rock. Amazing, this in only one half of the volcano. The other half is still just east of Los Angeles in Southern California.

Based on the previous movement and very sophisticated instruments and measurements, geologists can make some fairly reasonable predictions about what will happen to California and the rest of the continents in the future. However, like any prediction, the further out you go, the weaker the prediction becomes. Nevertheless, a view of the next 50 million years is pretty good. Robert S. Dietz of the USGS did some of the early work that validated plate tectonics on the ocean floor. In 1970, he wrote an interesting article in *Scientific American* in which he predicted the future of California. He said that 10 million years from now, Los Angeles will have moved up to the current Bay Area and in 50 million years, it will be part of Alaska. More recent work by geologists has validated the general lines of Dietz's projections.

Interestingly, some of the factors controlling the fate of the Bay Area are active at the Southern end of the San Andreas fault, east of San Diego. The process can be seen in the Gulf of California. A rift has formed. This highly geologically active area is beginning to break-off from Mexico. Eventually, it will open up into an inland sea as that part of California/Mexico moves northward. Christopher R. Scotese, a geologist at the University of Texas, Arlington, has calculated the results of future movements on the California coast. His website (www.scotese.com) contains projections of the Earth over the next 250 million years.

Some scientists believe that, in the next 8–10 million years, the Gulf of California will expand into the Walker Lane and extend northward to Lake Tahoe. The San Andreas fault has a long curve westward that is not favorable for movement. The more natural path is up Walker Lane. Walker Lane is on the California-Nevada border and runs north of Mount Lassen to Death Valley and beyond where it connects to the San Andreas fault system (Faulds et al., 2005). The belief is that the San Andreas fault will become much less active as the movement is taken up by this new system. If the scientists are correct, the rift that formed the Gulf of California will continue north on a path that is similar to that in Baja California. The path includes the Salton Sea, Mono Lake, Lake Tahoe, and Pyramid Lake and also a number of volcanoes.

While not everyone agrees with the new theory, there is growing evidence to support it. First, this little understood fault system accounts for 15–25% of the movement along the North American and Pacific plates. This is a surprising amount, considering that the San Andreas fault had been assumed to be the dominant interface between the plates. Second, earlier assumptions also could not account for the fact that the Baja Peninsula separated from the North American plate about 7 million years ago to form the Gulf of California. A chain of old volcanoes warmed the continental crust, causing it to soften and creating a series of weak spots that allowed the land to separate. Finally, about 13,000 years ago, Pyramid lake was part of an inland sea called Lake Lahontan. The lake is near a number of newly discovered faults, including the Pyramid Lake Fault, the Honey Lake Fault, and the Warm Springs Valley Fault. These developing faults might form part of the northern part of the new rift. The nascent rift in the Walker Lane provides an excellent opportunity to study a rift on land. Busby (2013) reviews just that. The fault system extends far beyond Lake Tahoe in the north. The rift in Walker Lane has moved northward as the Mendocino triple junction has also moved northward. That triple junction is where the Gorda, North American, and Pacific plates join at Cape Mendocino in Northern California. The Gorda plate is a subplate of the Farallon plate that broke-off as the Farallon was subducted under the North American plate (Fig. 10.3).

McCrory et al. (2009) modeled the subduction margins in Southern California and the slab windows that opened in that area. They cataloged the volcanic events that accompanied those slab windows from 28.5, 19, 12.5, and 10 million years ago. With these, they developed a model to describe the evolution of the continental margin. They then ran their model into the future to see what would happen. At 2 million years in the future, the bend in the San Andreas fault in Southern California becomes a problem. The Peninsula Ranges and the Sierras have been converging since 12.5 million years ago, but by 2 million years in the future, they will collide directly, and that pressure must be relieved somehow. McCrory et al. (2009) suggest that this stress could be relieved by new faults breaking through the lower Sierra Madre or by a shift of the movement between the plates from the San Andreas to the Walker Lane system.

So, there are two possible scenarios for the future of California and the Bay Area. In the first, more traditional view, the Pacific plate and that part of

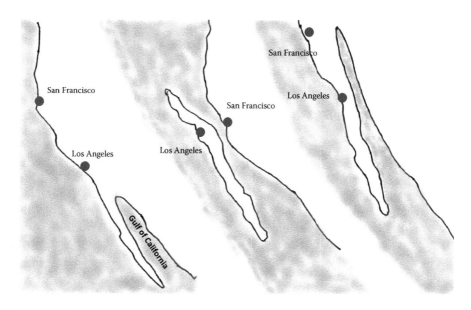

FIGURE 10.3   Two Models of the Future California. The forces that built the current California are still at work and eventually will tear it apart. Currently, the San Andreas fault runs from east of Los Angeles to just south of San Francisco, where it goes to sea (left). However, the Gulf of California is an increasingly active rift zone. Two models have been suggested for the future of the region. In the first, the part of California west of the San Andreas fault continues to move northward (middle). In the second, the stress moves to the Gulf of California, and the rift opens through the Walker Lane to north of Lake Tahoe (right).

California on that plate continue moving northward. In the second, the movement is transferred from the San Andreas fault to Walker Lane as that rift continues to extend northward. In this scenario, a much larger portion of California moves northward, including the entire Bay Area. In either case, California and the Bay Area will be unrecognizable.

REFERENCES

Agliardi F., Scuderi M.M., Fusi N., Collettini C. (2020) Slow-to-fast transition of giant creeping rockslides modulated by undrained loading in basal shear zones. *Nature Communications* 11: 1352.

Barnard P.L., Schoellhamer D.H., Jaffe B.E., McKee L.J. (2013) Sediment transport in the San Francisco Bay coastal system: An overview. *Marine Geology* 435: 3–17.

Becker N., Geschwind L.R. (2016) Earthquakes—2001–2015. *Science on a Sphere. National Oceanic and Atmospheric Administration.* http://sos.noaa.gov/Datasets/dataset.php?id=643.

Bever A.J., MacWilliams M.L., Wu F., Andes L., Conner C.S. (2014) Numerical modeling of sediment dispersal following dredge material placements to examine possible augmentation of the sediment supply to marshes and mudflats, San Francisco Bay, USA. PIANC World Congress San Francisco. http://www.southbayrestoration.org/documents/technical/PIANC2014Congress_SouthBayDredgePlacement_Bever_etal.pdf.

Busby C.J. (2013) Birth of a plate boundary at ca. 12 Ma in the Ancestral Cascades arc, Walker Lane belt of California and Nevada. *Geosphere* 9: 1147–1160.

Caldwell P.M., Bretherton C.S., Zelinka M.D., Klein S.A., Santer B.D., Sanderson B.M. (2014) Statistical significance of climate sensitivity predictors obtained by data mining. *Geophysical Research Letters* 41: 1803–1808. http://onlinelibrary.wiley.com/doi/10.1002/2014GL059205/full.

Carson M., Kohl A., Stammer D., Slangen A.B.A., Katsman C.A., van de Wal R.S.W., Church J., White N. (2016) Coastal sea level changes, observed and projected during the 20th and 21st century. *Climatic Change* 134: 269–281. http://link.springer.com/article/10.1007/s10584-015-1520-1.

Cazenave A., Le Cozannet G. (2014) Sea level rise and its coastal impacts. *Earth's Future* 2: 15–34. http://onlinelibrary.wiley.com/doi/10.1002/2013EF000188/full.

Chua V.P., Xu M. (2014) Impacts of sea-level rise on estuarine circulation: An idealized estuary and San Francisco Bay. *Journal of Marine Systems* 139: 58–67. http://www.sciencedirect.com/science/article/pii/S0924796314001419.

Cloern J.E., Knowles N., Brown L.R., Cayan D., Dettinger M.D., Morgan T.L., et al. (2011) Projected Evolution of California's San Francisco Bay-Delta-River System in a Century of Climate Change. *PLoS ONE* 6(9): e24465. https://doi.org/10.1371/journal.pone.0024465.

Cook B.I., Ault T.R., Smerdon J.E. (2015) Unprecedented 21st century drought risk in the American Southwest and Central Plains. *Science Advances* 1: e1400082.

Dettinger M.D., Ingram B.L. (2013) The coming megafloods. *Scientific American* 308: 64–71.

Diffenbaugh N.S., Swain D.L., Touma D. (2015) Anthropogenic warming has increased drought risk in California. *Proceedings of the National Academy of Sciences of the United States of America* 112: 3931–3936.

Faulds J.E., Henry C.D., Hinz N.H. (2005) Kinematics of the northern Walker Lane: An incipient transform fault along the Pacific–North American plate boundary. *Geology* 33: 505–508.

Gershunov A., Shulgina T., Clemesha R.E.S., Guirguis K., Pierce D.W., Dettinger M.D., Lavers D.A., Cayan D.R., Polade S.D., Kalansky J., Ralph F.M. (2019) Precipitation regime change in Western North America: The role of atmospheric rivers. *Scientific Reports* 9: 9944.

Goulden M.L., Bales R.C. (2019) California forest die-off linked to multi-year deep soil drying in 2012-2015 drought. *Nature Geoscience* 12: 632–637.

Goss M., Swain D.L., Abatzoglou J.T., Sarhadi A., Kolden C.A., Williams A.P., Diffenbaugh N.S. (2020) Climate change is increasing the likelihood of extreme autumn wildfire conditions across California. *Environmental Research Letters* 15: 094016.

Griffin D., Anchukaitis K.J. (2014) How unusual is the 2012–2014 California drought? *Geophysical Research Letters* 41: 2014GL062433.

Handwerger A.L., Huang M.-H., Fielding E.J., Booth A.M., Bürgmann R. (2019) A shift from drought to extreme rainfall drives a stable landslide to catastrophic failure. *Scientific Reports* 9: 1569.

Harou J.J., Medellín-Azuara J., Zhu T., Tanaka S.K., Lund J.R., Stine S., Olivares M.A., Marion W. Jenkins M.W. (2010) Economic consequences of optimized water management for a prolonged, severe drought in California. *Water Resources Research* 46: W05522.

Hays J.D., Imbrie J., Shackleton N.J. (1976) Variations in the Earth's orbit: Pacemaker of the Ice Ages. *Science* 194: 1121–1132.

Hirschfeld D., Hill K.E. (2017) Choosing a Future Shoreline for the San Francisco Bay: Strategic Coastal Adaptation Insights from Cost Estimation. *Journal of Marine Science and Engineering* 5(3): 42. https://doi.org/10.3390/jmse5030042.

Jelks H.J., Walsh S.J., Burkhead N.M., Contreras-Balderas S., Díaz-Pardo E., Hendrickson D.A., Lyons J., Mandrak N.E., McCormick F., Nelson J.S., Platania S.P., Porter B.A., Renaud C.B., Schmitter-Soto J.J., Taylor E.B., Warren, Jr M.L. (2008) Conservation status of imperiled North American freshwater and diadromous fishes. *Fisheries* 33(8): 372–407.

Klein A.-M., Boreux V., Fornoff F., Murepele A.-C., Pufal G. (2018) Relevance of wild and managed bees for human well-being. *Current Opinion in Insect Science* 26: 82–88.

Knowles N. (2010) Potential inundation due to rising sea levels in the San Francisco Bay Region. *San Francisco Estuary and Watershed Science* 8(1). https://escholarship. org/uc/item/8ck5h3qn.

Kopp R.E., Horton R.M., Little CM., Mitrovica J.X., Oppenheimer M., Rasmussen D.J., Strauss B.H., Tebaldi C. (2014) Probabilistic 21st and 22nd century sea-level projections at a global network of tide-gauge sites. *Earth's Future* 2: 383–406. http://onlinelibrary.wiley.com/doi/10.1002/2014EF000239/abstract.

Maasch K.A. (1997) What triggers ice ages? *Nova.* http://www.pbs.org/wgbh/nova/earth/ cause-ice-age.html.

Maslin M. (2016) Forty years of linking orbits to ice ages. *Nature* 540: 208–209.

Maslin M. (2020) Tying celestial mechanics to Earth's ice ages. *Physics Today* 73: 48–53.

McCrory P.A., Wilson D.S., Stanley R.G. (2009) Continuing evolution of the Pacific–Juan de Fuca–North America slab window system—A trench–ridge–transform example from the Pacific Rim. *Tectonophysics* 464: 30–42.

Meehl G.A., Arblaster J.M., Marsh D.R. (2013) Could a future "Grand Solar Minimum" like the Manunder Minimum stop global warming? *Geophysical Research Letters* 40: 1789–1793.

Mitchell, R.N., Kilian, T.M., Evans, D.A.D. (2012) Supercontinent cycles and the calculation of absolute palaeolongitude in deep time. *Nature* 482: 208–212.

Molnar J.L., Gamboa R.L., Revenga C., Spalding M.D. (2008) Assessing the global threat of invasive species to marine biodiversity. *Frontiers in Ecology and Evolution* 6: 485–492.

Moreno-Mateos D., Power M.E., Comin F.A., Yockteng R. (2012) Structural and functional loss in restored wetland ecosystems. *PLoS Biol.* https://doi.org/10.1371/journal.pbio.1001247.

O'Gorman P.A. (2015) Precipitation extremes under climate change. *Current Climate Change Report* 1: 49–59.

Putirka K.D., Busby C.J. (2011) Introduction: Origin and evolution of the Sierra Nevada and Walker Lane. *Geosphere* 7: 1269–1272.

Rosenberg K.V., Dokter A.M., Blancher P.J., Sauer J.R., Smith A.C., Smith P.A., Stanton J.C., Panjabi A., Helft L., Parr M., Marra P.P. (2019) Decline of the North American avifauna. *Science* 366: 120–124.

Ruckert K.L., Oddo P.C., Keller K. (2017) Impacts of representing sea-level rise uncertainty on future flood risks: An example from San Francisco Bay. *PLoS ONE* 12(3): e0174666.

Stephens S.A., Bell R.G., Lawrence J. (2017) Applying principles of uncertainty within coastal hazard assessments to better support coastal adaptation. *Journal of Marine Science and Engineering* 5: 40. doi:10.3390/jmse5030040.

Stralberg D., Brennan M., Callaway J.C., Wood J.K., Schile L.M., Jongsomjit D., Kelly M., Parker V.T., Crooks S. (2011) Evaluating tidal marsh sustainability in the face of sea-level rise: A hybrid modeling approach applied to San Francisco Bay. *PLoS ONE* 6(11): e27388. http://journals.plos.org/plosone/article?id=10.1371/journal.pone.0027388.

Suddeth R.J., Mount J., Lund J.R. (2010) Levee decisions and sustainability for the Sacramento-San Joaquin Delta. *San Francisco Estuary and Watershed Science* 8(2). https://doi.org/10.15447/sfews.2010v8iss2art3.

Swain D.L., Langenbrunner B., Neelin D., Hall A. (2018) Increasing precipitation volatility in twenty-first century California. *Nature Climate Change* 8: 427–433.

Takekawa J.Y., Throne K.M., Buffington K.J., Spragens K.A., Swanson K.M., Drexler J.Z., Schoellhamer D.H., Overton C.T., Casazza M.L. (2013) Final report for sea-level rise response modeling for San Francisco Bay estuary tidal marshes. Open-File Report 2013-1081. US Geologic Survey. https://pubs.er.usgs.gov/publication/ofr20131081.

Tilman D., May R., Lehman C.L., Nowak M.A. (1994) Habitat destruction and the extinction debt. *Nature* 371: 65–66.

Wake D.B., Vredenburg V.T. (2008) Are we in the midst of the sixth mass extinction? A view from the world of amphibians. *Proceedings of the National Academy of Sciences of the United States of America* 105: 11466–11473.

Wan H.Y., Cushman S.A., Ganey J.L. (2019) Recent and projected future wildfire trends across the ranges of three spotted owl subspecies under climate change. *Frontiers in Ecology and Evolution* 7: 37. https://doi.org/10.3389/fevo.2019.00037.

Index